The Mocking Memes

A Basis for Automated Intelligence

by

Evan Louis Sheehan

authorHOUSE™

1663 LIBERTY DRIVE, SUITE 200
BLOOMINGTON, INDIANA 47403
(800) 839-8640
WWW.AUTHORHOUSE.COM

AuthorHouse™
1663 Liberty Drive, Suite 200
Bloomington, IN 47403
www.authorhouse.com
Phone: 1-800-839-8640

AuthorHouse™ UK Ltd.
500 Avebury Boulevard
Central Milton Keynes, MK9 2BE
www.authorhouse.co.uk
Phone: 08001974150

First published by AuthorHouse 10/10/2006
ISBN: 1-4259-6160-6 (sc)

Printed in the United States of America
Bloomington, Indiana

This book is printed on acid-free paper.

Table of Contents

Preface

If you've ever been face to face with a hovering hummingbird, or eye to eye with a trusting ferret, ... if you've ever handed a graham cracker to a wild raccoon, ... had a kitten curl up in your lap, ... watched a spider catch a fly, ... or, if you've ever seen a kingfisher dive into water and emerge with a fish, ... well, then perhaps you can understand my rapture with life, its origins and its natural motivations. Perhaps you can appreciate my obsession with understanding evolution.

I've been interested in evolution for as long as I can remember. It has always seemed evident to me that the implications of evolution, regarding such things as free will, morality and intelligence, are far more dramatic than we generally tend to acknowledge. But, as a teenager, only occasionally did I ever dwell on those dramatic implications. Instead, like most teenage boys, my thinking primarily revolved around my friends and acquaintances, especially girls. And, in a precocious manner, I happily justified my fetish for females as being completely consistent with evolutionary principles.

As an undergraduate in college, my naïve interest in evolution began to mature into an amateurish fascination. It happened for me after reading Carl Sagan's book *The Dragons of Eden* (1977). Ever since, I've been trying to reconcile evolutionary concepts with human behavior and intellectual capability. But my pursuit of a deeper evolutionary comprehension of life remained little more than a hobby, for decades, simply because I recognized some insurmountable hurdles standing between me and a complete understanding – there were some critical pieces missing from my conceptualization of the puzzle.

Not until several years ago, when I read Richard Dawkins' revolutionary book *The Selfish Gene* (1976), did my fascination with evolution quickly evolve into an obsession. Suddenly, the completion of the puzzle was within reach – there were no more insurmountable hurdles, no more missing pieces. I finally understood how love, friendship, art, culture, technology, intelligence, consciousness, emotions and even morality were able to emerge from purely physical processes. It was a revelation for me.

I felt a need to assemble the puzzle into a coherent picture of human life. And so I started to write, and think, and write, and think some more. It is amazing how the act of writing inspires thinking, which in turn inspires more writing, which requires more thinking, in something of a repetitive evolutionary process. I knew the act of writing would force my thinking to become more precise. Indeed, my desire for precision and coherence of thought was the primary reason for my starting to write. If you ever want to clarify your thoughts on a particular subject, just write a book about it.

My original intent was to organize my thoughts by writing a book that would reconcile evolution with … well, just about everything. I wanted to leave as a legacy to my children and their descendants a well-considered perspective – a scientific perspective – on the meaning of life. But, in the process of writing, I decided there were too many important ideas to fit into one book. So, the growing text, which I had originally intended to title *The Laughing Genes and the Mocking Memes*, split into two books. The first, titled *The Laughing Genes*, would focus on reconciling evolution with morality. I published that book in 2005. This is the second book – a companion book – which is intended to fully reconcile evolution with human intelligence.

I seek here to present a theory, completely consistent with evolution, on how the brain automatically organizes itself and thereby becomes intelligent. I'll necessarily draw on many of the concepts developed in my previous book, regarding human will, emotions, consciousness, and the fundamentals of memetic evolution. And, since I want this book to stand on its own, I'll have to repeat some of the concepts previously developed. I won't spend much time on them, and I'll try to come at them from a slightly different angle. But, I encourage you to read *The Laughing Genes*, as a companion to this book, in order to gain a deeper understanding of any critical concepts found to be lacking in support.

You and I, separated by mere space and time, will cooperatively explore the subject of memetic intelligence. We cooperate by virtue of my willingness to *propose* and your willingness to *evaluate*, with respect to this specific and very abstract subject. Those complementary characteristics of our respective wills fit together like two perfectly mating puzzle pieces, thereby allowing a cooperative synergy to emerge. The fundamental concept of *cooperation*, applied at many different levels, plays a significant role in the understanding of intelligence, just as it did in my first book regarding the understanding of morality. I endeavor to show that the synergy emerging from cooperation is the very essence of all sorts of intelligence.

Cooperation always involves complementary interfaces between cooperating components. It is the fundamental job of intelligence to identify which environmental elements have complementary interfaces, so that they may be synergistically combined and used for benefit. Biological evolution finds complementary interfaces randomly. But human brains automatically sort concepts by their abstract interfaces, in something of a multi-dimensional conceptual space, so that complementary interfaces can be more easily identified. Human cognition achieves its intelligence through an automatic process that is something like algorithmically sorting the pieces of a jigsaw puzzle by their shapes or colors before searching for a particular piece. Proper sorting of neural connection patterns enables much faster searching by evolving neural firing patterns.

In addition to being a book about intelligence, this is also a book about *memes*. Through cooperation with many readers I hope to breathe new life into the flagging field of memetics. In addition to rekindling interest, I intend to establish the basis for a true science of memes, about which only a handful of books have ever been written. I hold a strong conviction that the concept of memes, properly understood, has the potential to explain all forms of intelligence, including even the intelligent mind of God as it applies to the creation of life and even the universe. The concept of memes and the concept of synergistic cooperation are fundamental characteristics of nature that, when considered together, allow us to unify everything in the spirit of Edward O. Wilson's book *Consilience: The Unity of Knowledge* (1998).

I'll present a theory of human intelligence that is built from the joining of several critical and complementary concepts. While viewing the world from Richard Dawkins' replicator-centric perspective, I'll combine his concept of memes with Jeff Hawkins' proposed neural architecture, so as to fully realize Gerald Edelman's theory of *Neural Darwinism*.

Edelman, a Nobel Prize-winning doctor and scientist, essentially modeled the workings of the human brain as, itself, something of an evolutionary process. There has been a lot of confusion surrounding his book *Neural Darwinism* (1987), and because of that confusion his theories have never really gained acceptance. I hope to bring clarity by presenting some of his well-founded ideas in the light of a slightly different context – a memetic context.

The memetic perspective on intelligence is a very troubling concept for many people. It seems to strip the mind of all volitional control. I find that even scientists tend to have a knee-jerk reaction against it. Hence, I have had a difficult time getting knowledgeable people to invest the effort to read about it. And so I am extremely grateful to Kirk Tennent and Doug Smith for their willingness to wade through early versions of this text. Their words of encouragement have inspired me greatly. Finally, I want to thank my family – my dear wife, my wonderful kids, and my loving parents – for tolerating my development of a cognitive theory that so many people find to be distasteful.

I fully admit that my family, all my acquaintances and all my experiences have made me what I am. I completely accept that I am solely a product of my heredity and environment – inherited genes and acquired memes. In such a worldview, all my decision-making has been, and will forever be, mostly, if not completely, a matter of destiny rather than free will. Despite such a fatalistic perspective, I find myself to be a very content individual. My scientific view of life is indeed radically different from most everyone else's, yet, for me, it is extremely satisfying in its obvious truthfulness. We don't diminish the magnificence of humanity at all by accepting the replicator-centric view of life. And, as I explained in my previous book, we needn't change our ethical values either. I happen to believe the world would be a much better place if everyone held the same scientific perspective on the purpose of life and the memetic nature of automated intelligence that I have embraced. Please, allow me to share it with you.

Chapter 1. **Evolving Patterns**

There is no source of creativity anywhere in the universe other than the process of evolution. That is the simple premise underlying this book. Indeed, I shall be describing all sorts of creative intelligence, human and otherwise, as various styles of evolution operating on various patterns of things, such as patterns of DNA, patterns of human behavior and patterns of neural firings. My fundamental thesis of human intelligence, then, is this: Just as evolving patterns of DNA have resulted in the 'Intelligent Design' of biological humans, so do evolving patterns of neural activity automatically give rise to intelligent thoughts and rational human behaviors.

Before we can understand how an evolutionary process can apply to patterns of neural activity, we first have to understand the fundamental properties of all *patterns*. What is it that constitutes a pattern? What constitutes a class of patterns? How do patterns replicate? How do multiple patterns combine in ways that enable them to achieve synergy? And how are patterns likely to degrade and mutate over time? By the end of the book, it will become clear that the fundamentals of human intelligence are a lot like the fundamentals of biological evolution. They both involve iterated operations on patterns, and they both produce amazing results, given enough time. Granted, there are a few important differences, but, underlying both, there appears to be a simple process of iterated replication, variation and differential selection applied to many sorts of patterns.

The incremental advance of intelligence in the evolutionary history of species, leading up to humans, strongly suggests that intelligence is an automatic and algorithmic process, constantly refined by evolution. But then, what is it that accounts for the enormous leap to human intelligence? I am not the first to suggest that the difference is accounted for by the cognitive facilitation of rapidly evolving cultural and technological replicators, defined by Richard Dawkins as *memes* (rhymes with *dreams*).

I'll slightly redefine memes in a way that portrays them as the pivotal elements in a coherent theory of mind. The theory reconciles all sorts of creativity, natural and human, as processes of evolving patterns. Memes take on a more universal role, best characterized as the fundamental elements of intelligence – the currency of thought, for all sorts of intelligence.

Prepare yourself for a completely different perspective on what it means to be intelligent – a memetic perspective. The automatic and algorithmic nature of that perspective will enable us to design machines that powerfully augment our human creative abilities and eventually even exceed them. So, before we get into the nitty-gritty of precisely defining memes and patterns, let us contemplate the serious need to do so, by considering the likely future of computational intelligence.

Finding a basis for automated intelligence seems to be the next logical step for our species in the natural progression of life. Indeed, I'll endeavor to show that there is a natural direction to be expected in the progression of all life, everywhere in the universe. I call it the *arrow of evolution*. Such a statistical destiny guarantees that all instances of planetary life will eventually become intelligent, simply because intelligence yields survival value in absolutely any environment.

Once intelligent life reaches a critical level, characterized by an ability to understand the nature of its own intelligence, it will inevitability undergo a revolutionary transformation. The revolution occurs when life understands its own intelligence well enough to begin developing tools that further enhance its intelligence. Just as the development of the abacus, the language of mathematics, the slide rule and the calculator have extended the functionality of the human brain in the past, computer technology will likely extend our future powers of reasoning in such a dramatic fashion that we have trouble even imagining it. The memetic perspective will enable us to see exactly why intelligence breeds more intelligence at a rate that accelerates over time.

The revolutionary transformation is complete when intelligent beings construct machines that are able to think as creatively as themselves. All evidence indicates that earthly life is poised at this amazing threshold.

It is easy to dismiss this view as sensationalistic. But, historical trends clearly show that the advancement of computing power is indeed accelerating. We are at the knee of an up-turning curve, and if historical trends continue into the future, we will see the development of stupefying computing power in just the next 50 years. Ray Kurzweil, a brilliant inventor and entrepreneur, lays out the rationally expected future in his book *The Age of Spiritual Machines: When Computers Exceed Human Intelligence* (1999). He compares a modern $1000 computer to the computing power of an insect. By the year 2010, a similarly affordable computer will be comparable to the computing power in the brain of a mouse. By 2030, the same value will buy more computing power than is in a human brain. And, by 2060, a single, cheap computer will possess more computing power than is in all existing human brains put together. Every person will own several of these ultra-powerful computers and the Pentagon will own a computer a million times more powerful than even that – if trends continue.

I am inclined to believe that, within fifty years, machines will communicate with us in our natural languages. Perhaps that threshold is only twenty years away, or less. Whenever it occurs, computers will then digest and organize the mountains of information in all of our books. Just from reading our books, they'll identify important scientific relationships that never occurred to us. Understanding, in all areas of technology, will blossom unimaginably.

Even if trends fail to maintain the super-frenetic pace of acceleration, I believe it is safe to say we are in for some big surprises in the next fifty years. Thus, I am highly motivated to try to understand the driving force behind evolving intelligence, simply because I want to meaningfully participate in, and benefit from, the upcoming revolution.

Intelligent machines will not result from some completely random sort of event, such as a bolt-from-the-blue 'eureka' moment. No, just as all important discoveries inevitably spring from an ever growing foundation of knowledge, computational intelligence is statistically destined to emerge from the confluence of several more-primitive understandings. Computational intelligence naturally and inevitably evolves by combining complementary patterns of knowledge into new and better patterns of knowledge.

A full comprehension of evolution is key to understanding intelligence, not just because evolution created all known intelligence, but more importantly, because evidence shows that all intelligent systems, including human brains, use various styles of evolutionary algorithms as their basic means for achieving intelligence.

All current computational methods of artificial intelligence can be classified as subsets of evolutionary algorithms. But, instead of searching by random mutation, as nature seems to do, computers tend to search by a more thorough and systematic style of mutation as they evaluate and select within a simulated environment. Yet, despite the obvious similarities between artificially intelligent systems and the natural intelligence of biological evolution, few experts in the area of artificial intelligence have any sort of background that would allow them to fully understand biological evolution.

On this point – that creativity always involves a process of evolution – rests the unique value of this book. Of the many books on cognition and intelligence, few even mention the *memetic* class of evolving patterns that, I'll argue, are responsible for all our human thoughts. This book explores them in depth.

Because this book so radically departs from traditional views on human cognition, its journey of explanation must start at the most fundamental levels. From the careful definition of *memes* and the mundane consideration of *patterns*, the journey necessarily requires us to wade through a slow and methodical analysis of various evolutionary processes operating on many sorts of patterns. A lot of groundwork needs to be laid. A lot of traditional thinking needs to be methodically re-oriented. Layer by layer, we must range across many diverse topics in an awkward process of building the proper foundation for a memetic perspective on intelligence. Unfortunately, these prerequisite fundamentals take up at least the first half of the book.

The ultimate goal is to understand exactly what the human brain has been evolutionarily designed to do. And the payoff is considerable. We will be able to identify an algorithmic mapping between human perception and somewhat specific patterns of neural flow within the human brain. From there, we will be in a position to identify the mechanism by which those patterns of neural activity are able to rapidly evolve so as to produce human thought and consequent behaviors.

Amazingly, we will come to understand human cognition as an automatic process of evolving memetic patterns, all happening within a relatively simple and highly redundant neural architecture. The human brain is thus explained in terms of its architectural ability to automatically capture, express, and evolve memetic patterns of relationships, which correspond to our thoughts. A machine architecture for duplicating the functionality of the brain will become somewhat apparent, but not yet easily implemented.

Interspersed among discussions of how to map memes onto neural structures are some philosophical discussions that serve as metaphors for understanding, and also, as palatable bridges toward the emotional acceptance of a mechanistic sort of human mind. I'm afraid the philosophy of morality is hopelessly entangled with the process of intelligence, as evidenced by the simple and profound question: Is it intelligent to be moral? I'll clearly show that the answer is 'yes', when morality is properly defined. The book concludes with a philosophical view on human purpose, Gaia, Intelligent Design, and how it seems to be the case that even the entire universe appears to have been intelligently created by an evolutionary sort of process.

The Essence of Memes

Richard Dawkins certainly started something big when he wrote about *memes* in his landmark book *The Selfish Gene* (first published in 1976, and subsequently re-released in 1989 with two additional chapters). I doubt that he even realizes how big an idea it will ultimately turn out to be. I make that bold statement only after reaching a general understanding of how physical brains and patterns of memes combine to create intelligent minds. I claim that, when precisely defined, memes are responsible for all interesting aspects of all life, especially human life. Indeed, they account for the great difference between us and the rest of the living world.

I'll define memes in a way that portrays them as the essence of all intelligence and creativity. When properly understood, they not only enable evolution of various sorts, they are also the products of evolution. And when the products of evolution are fed back into the system as better enablers of evolution, they are sometimes able to form whole new loops of evolutionary feedback on top of already existing evolutionary loops. They form advanced styles of evolution. Thus, the process of evolution itself evolves, and memes are the facilitators of that accelerating process.

The concept of memes proposed by Dawkins grew out of his recognition that biological evolution has three primary requirements, all of which also tend to be satisfied in the process of cultural development. The three ingredients for evolution of genes are: replication, mutation, and differential selection. Perhaps, speculates Dawkins, it is not a coincidence that culture evolves similarly to the way biology evolves. After all, cultural behaviors do indeed seem to get replicated from person to person, are sometimes mutated, and are differentially selected for performance. So, Dawkins defines memes in terms of those replicating cultural behaviors.

As I speak of memes, I'll be thinking of Dawkins' term in a much broader sense than he and others tend to use it. And so, I need to describe precisely how my definition is different from his. Let us first review Dawkins' definition for memes.

Dawkins describes the meme as a *unit of imitation*, primarily responsible for the propagation of culture. As children automatically imitate the behaviors of their parents, culture propagates. Dawkins loosely defines his classification of memes by citing the following examples: "tunes, ideas, catch phrases, clothes fashions, ways of making pots or of building arches." While these examples give us a general idea of what memes are, we need to be a bit more specific in our definition, by explicitly defining the boundaries of the class.

Many memeticists have tried to pin down the definition of a meme by limiting its scope to an explanation of culture only. But I find that my inability to decide exactly what does and does not fit within the boundaries of culture leads me in the opposite direction. Is language part of culture? What about technology? They both obviously evolve in ways similar to how culture evolves. So, instead of limiting the scope of memes, I choose to broaden it by identifying the essence underlying *all* classes of things that evolve. Such an essence allows us to define many things in terms of memes, including culture, and, more interestingly, all forms of creativity and intelligence. Only by broadening the definition of memes can we gain a better understanding of how life initially became intelligent, and thereby reveal exactly how life will soon become even far more intelligent.

I define memes to include *every sort of pattern that serves as a template for its own replication*. Certainly, all the instances cited by Dawkins, including all aspects of culture, fit this description of memes. But I can think of other things beyond what Dawkins might include. For example, the pattern of atoms on the face of a growing crystal is a meme, because it serves as a template for its own replication through the formation of subsequent layers. And patterns of DNA also fall into the class of patterns that serve as templates for their own replication.

I'll later make the case that even patterns of neural firings within a brain are memes, because they establish templates for future patterns of firings. Although, to be perfectly precise, firing patterns are first translated to neural connection patterns (synaptic weightings) and then later translated back again into neural firing patterns similar to the original firing patterns from which the neural connection patterns were established. So, before we make the claim that firing patterns replicate themselves, we need to better understand the act of translation.

It is crucial to recognize that pattern translation absolutely does involve a form of pattern replication, even though the replication happens abstractly from one 'substrate' to another. Consider the act of translating a spoken sentence either into a tape-recording or into text. In both cases, the meaning of the sentence is unaffected and the original pattern can be loosely reconstituted by either playing back the tape or reading the text aloud. Recognize that, in the substrate of text, the translated pattern is completely different from the pattern of audio waves in the original spoken sentence, and any reverse-translation is likely to involve some loss of fidelity. Of course, the act of reading aloud, itself, involves many sorts of translations.

The accuracy and reversibility of a translation depends entirely on the substrates involved. All translations of patterns, from one substrate to another, happen strictly by the laws of nature. And nature has arranged for some sorts of translations to be very accurate and reversible, but not all. For patterns that undergo completely accurate and reversible translation, the abstract characteristics of the original pattern must indeed remain in the translated pattern, somehow, and the original pattern is thereby replicated in a very real sense. We absolutely must deal with acts of translation as means for pattern replication, because circular translations can produce a true act of replication.

Circular chains of translations seem to play a big part in the replication of cultural memes. Consider the cycle of translations starting from patterns of behavior (performed by others), which are automatically translated to patterns of light, which are then translated to patterns of neural photoreceptor firings in the back of one's eye, which ultimately get translated into patterns of synaptic connections, which later complete the cycle as they get translated back to neural firing patterns, causing muscle contraction sequences, giving rise to patterns of behavior. This cycle of translations forms the general mechanism by which cultural memes get themselves replicated. It seems evolution has discovered various ways of replicating patterns through complex combinations of automatic translations.

We now find ourselves having to broaden our definition of memes even further to include all patterns that serve as templates for their own replication *or accurate translation*. In the interest of keeping things simple we won't need to deal much with translation. So, from here on out, we'll simply focus on the ultimate effects of replication, without worrying about any involved translations.

Dawkins might accuse me of having gone too far with his concept of memes. Indeed, he stated in his book *Unweaving the Rainbow* (1998) that: "It is a matter of dispute whether the resemblance between gene and meme is good scientific poetry or bad. On balance, I still think it is good, although if you look the word up on the worldwide web you'll find plenty of enthusiasts getting carried away and going too far." While I am sensitive to his concerns, I hold a deep conviction, supported by reason, that memes, when they are properly defined, are at the heart of all intelligence, even the intelligence of biological design. So, I hope Dawkins will eventually agree that I haven't gone to far with his concept, and that in fact he hadn't gone far enough.

There is inherent ambiguity in Dawkins' definition of memes. The ambiguity comes from the lack of a concise definition for the word 'imitate'. For example, Dawkins cites religious beliefs as a powerful and devious set of memes. But what exactly is a belief? Does it fit his definition of a meme as a unit of imitation? Can a belief be imitated? The answer is not completely clear, because imitation usually implies a behavioral action of some sort. But then, how about my definition? Can a belief be a pattern that serves as a template for its own replication or translation? I have to admit to some ambiguity in what it is that constitutes a pattern. So, the first thing we need to do is to define exactly what we mean when we speak of patterns. We'll do that soon. And, eventually, we'll see that beliefs are clearly defined by relationships between patterns of neural activity within the brain. Hence, we'll find that beliefs can indeed serve as templates for their own replication or translation whenever they are, for example, reconsidered or communicated to others.

Dawkins makes the point that genes, by evolving brains, have created a 'nutrient-rich soup' in which the new style of replicators – the memes – are able to evolve. We'll find this to be a general theme whereby evolution itself is able to evolve. We'll discover that lower-level memes cooperate so as to facilitate higher-level memes. And those higher-level memes cooperate so as to facilitate yet higher ones, thereby building a hierarchy of memes, in what may be considered a general principle of evolving evolution. At every level, the memes at that level create something of a 'nutrient-rich soup' for the creation of yet higher-level memes. And it matters not what the substrate of patterns is at any particular level. Patterns are patterns, whether they are composed of DNA, neural firings, behaviors, or even plain old cardboard.

By defining memes to include all patterns that serve as templates for their own replication or translation, I hope eventually to express intelligence in terms of algorithms for manipulating memes. Realize, however, that such a broad definition includes genes as a subset of memes. As I previously mentioned, genes certainly do serve as templates for their own replication. Indeed, my definition for memes recognizes genes as the prototypical memes.

Having precisely defined exactly what I mean when I use the term 'memes', I'll backtrack a bit by subscribing to its more common usage. So that I am consistent with other authors, I'll often use the term to describe all replicating patterns that serve as templates for replication or translation, *other than genes*. I'll just carve out the section of memes that is responsible for the phenotypic construction of all biological earthly life. I'll continue to refer to those replicators separately as 'genes', even though there is no fundamental functional distinction between them and the broader class of 'memes', as I've defined memes.

This leaves us with a much broader set of replicators underlying the propagation of culture. That set of replicating patterns necessarily includes things beyond what can be imitated, such as patterns of neural firings inside human brains. But why shouldn't we include those patterns of cognition as parts of culture when indeed cognitive processes are critical to how culture progresses. If our goal is to find the essence of intelligence, then we shouldn't be surprised to find that it underlies all intelligent processes, including the design of culture, the design of technology and even the design of biology. While I will usually refer to memes as separate from genes, we must keep in mind that they are fundamentally the same – patterns that serve as templates for their own replication or translation.

Why should we bother with these subtle refinements of definition? What real difference does it make? Well, it is all a matter of perspective. We may look at intelligence from several vantage points, and we must first come to an understanding as to which of these perspectives is best. Dawkins likes to use the Necker Cube[1] to illustrate how there can be multiple ways of looking at things. The different orientations of a Necker Cube correspond to the different meanings that can be taken from different perspectives, none being better than the others.

Dawkins' implication is that a gene's-eye-view of life is just as valid as an individual's perspective on life. According to Dawkins, it is just as appropriate to say that our genes use us humans to get themselves replicated as it is to say that we humans use our genes to make babies. I believe Dawkins' analogy here falls way short of the truth. The gene's-eye-view is a much *better* perspective on biological life. Only that perspective can enable us to understand human will, human morality and the purpose of life. So, I'll analogize the situation a bit differently.

Imagine two completely different perspectives, held by two naïve examiners, on a television set – one viewing it from the front and the other from the back. Is either viewpoint better than the other for trying to understand what the television does? Of course, only the frontal perspective allows an examiner to understand the significance of the television and all its components. If we are to understand the significance of intelligence, we must consider it from the better perspective. Rather than assuming that we humans use our intelligence for our own benefit of pursuing happiness, let us instead assume that the purpose of our human intelligence is for the perpetuation of our defining replicators – our genes and our memes. Only that proper perspective will enable us to understand the nature of intelligence, and the significance of the components inside our brains.

Before we are able to properly contemplate a memetic perspective, we need to dramatically change our views on life and its behaviors. We must understand the philosophical concept of determinism, and how it seems to apply to every physical process in the natural universe, including processes of the human brain. Such an understanding will then demand a brand new approach to thinking about intelligence, starting with the fundamentals of natural pattern formation and the deterministic laws by which patterns are forced to change.

Determinism

When Dawkins speaks of memes, he is referring to aspects of culture that propagate from person to person in something of an automatic fashion. Such a recasting of cultural elements, as autonomously replicating entities, entitles him to view them as having something of a propagation agenda of their very own. However, as he is careful to point out, he does not mean to imply that memes volitionally intend to seek ubiquity, but rather, he only means to imply that they appear to have the intention of spreading themselves, in the same metaphorically 'selfish' manner as genes. But in a world governed by natural laws that are mostly deterministic, even the intentions of human minds reduce to the mechanistic processes of human brains. That is, we humans don't choose volitionally, but rather, automatically.

Here lies a source of confusion between my definition for memes and the definition implied by many other authors. I see the mind as a deterministic device that processes ideas in a completely algorithmic manner. Daniel Dennett (1995, p.346) characterizes my view beautifully, but with contempt. He says: "I don't know about you, but I am not initially attracted by the idea of my brain as a sort of dung heap in which the larvae of other people's ideas renew themselves, before sending out copies of themselves in an informational diaspora. It does seem to rob my mind of its importance as both author and critic. Who's in charge, according to this vision – we or our memes?"

I was somewhat critical of Dennett in my previous book for having laid out so much strong evidence against free will, and then drawing the opposite conclusion: that we humans do have some sort of volitional agency. In the end, he allows his emotions to dominate his rational thinking. And, indeed, most authors similarly seem to hold out the prospect for some sort of free will, even in the face of overwhelming evidence against it. This is a pivotal point when it comes to discussions of memes. For, if indeed humans have the freedom of will to whimsically accept some memes and reject others, then we can't say that memes propagate by automatic processes, but rather, we are forced to characterize the propagation of memes as dependent on some undefinable cognitive factor commonly referred to as 'volition' or 'free will'. We might as well speak of a soul or a spirit.

Do children mimic the behaviors of their parents because they volitionally choose to, or because they are programmed by their genes to do so? Do teenagers mimic the behaviors of their peers on the basis of whim and fancy, or is there a more fundamental urge that automatically results from heredity and environment? We have to be absolutely clear on this topic. It is fundamental to all my arguments. Volition implies the springing into existence, from nothingness, of an uncaused event. Supposing the possibility of a volitional act puts a rip in the intricate fabric of causality, and forever prevents us from weaving a coherent description of intelligence. There can be no such thing as volition.

Science is built on the idea that everything in the universe operates according to very strict laws of nature. The quest of science is to discover those laws and express them mathematically. We should be grateful to God – whatever that is – for having put us in a universe that is so reliable in the nature of its unfolding. That reliability has allowed us to predict the future of simple systems with a high degree of certainty, and of even complex systems, albeit with a lower degree of confidence. For instance, we can be very certain about the simple behaviors of the ocean tides for years into the future, because they depend entirely on the simple and predictable behavior of the moon. And we can have some degree of confidence in our predictions of the complex behavior of the local weather over the next day or two.

The evolutionary purpose of intelligence is to use the predictability of nature for advantage toward survival. By learning about relationships observed in the past, an intelligent being can logically expect certain things to happen in the future. Such an ability to make predictions depends entirely on the memory of past experiences, and on the rational expectation that things will happen in the future as they did in the past. In other words, the ability to predict the future depends on the universe being deterministic. Future states will be determined through strict natural laws from present states, and present states have been determined through the laws of nature from previous states. All scientific evidence supports the idea that this condition of determinism indeed exists, at least to an extremely high degree.

Notice that all the neurons of the brain act by the very strict rules of physics. Researchers routinely stimulate various human sensors and then study the resulting neural responses. The effects are highly repeatable and the related biological systems seem to be very reliable in their functionality. Electronic signals propagate down nerve fibers with statistical regularity. It is nice to know that one can count on the reliable operation of one's brain without even thinking about it. It just happens. I find comfort in knowing that my mind keeps on working all the time. There's nothing I can do about it. It just keeps on working and working. But, an uncomfortable consequence of this view is that I don't control the flow of electronic impulses within my brain, the natural laws of physics do. If my thoughts and feelings result from the electronic impulses within my brain, then the laws of physics control my thoughts and feelings.

Most people, including many scientists, reject the idea that human minds are deterministic, only because they find it unappealing. They are perfectly willing to accept that the human heart reliably beats about two billion times in a human's life, completely as a result of the physical laws of nature operating on biological elements. But they are unwilling to believe that the human brain similarly obeys the laws of nature. Indeed, there have been many attempts at locating some transcendent component within the human brain that might act as a possible source of free will. Let us briefly explore some of them.

Chaos theory has shown that very complex behaviors can result from the interactions of certain types of simple systems. Edward Lorenz showed in a landmark paper titled "Deterministic Nonperiodic Flow" (1963) that the mathematics of certain three-variable, nonlinear systems of differential equations produce wildly erratic behaviors, seemingly out of nowhere. The unpredictable nature of certain chaotic systems has been cited by some philosophers as a potential source for free will in a deterministic universe. But they couldn't be more wrong.

Just the simple fact that these emergent phenomena can be produced by mathematical simulation tells us that they can be mathematically *re-produced*. We can be guaranteed that any mathematical reproduction of a wildly unpredictable behavior must happen exactly the same way every time we run the mathematical simulation. In a book called *Sync* (2003, p.188), Steven Strogatz writes: "The equations themselves contain no noise or randomness or other sources of uncertainty. Furthermore, if you solve the equations on a computer, using the same starting values for all the variables, the predicted outcome will be the same every time." So, if we can computationally simulate a chaotic system faster than real-time, then we can predict what that system will do in the future. The behavior can no longer be considered unpredictable. And it cannot be a source of volition. It is clearly determined and completely volitionless.

Even though chaotic systems can exhibit unusual behaviors, nothing in chaos theory suggests that those behaviors are not deterministically produced. We may not be able to predict the course of a naturally chaotic system, simply because its behavior depends so sensitively on the initial conditions of the system. But it is merely the case that we don't have the precision in our equipment to measure the initial conditions accurately enough. The bottom line of chaos theory is that certain types of systems can exhibit wild behaviors without violating determinism. In regard to such a system, James Gleick says in his book *Chaos* (1987, p.251) "The system is deterministic, but you can't say what it's going to do next." I disagree. In fact, you *can* say what it's going to do next if you have another instance of the very same deterministic system running slightly ahead of it.

Our ability to predict the future appears to be limited by the precision with which we can measure the current state of the world. This limitation of measurement precision (discovered and mathematically quantified by

Werner Heisenberg) is known as *Heisenberg's uncertainty principle.* The theory postulates a sort of 'fuzziness' related to the properties of particles at sub-atomic scales. These very small particles operate by a weird set of rules, known as quantum mechanics, that even allow things to randomly pop in and out of existence. Some have looked to these quantum-level effects as a source for free will (see Penrose, 1989). But they can only be a source for randomness, which is completely undesirable if we want to think rationally and reliably.

Whether we want it or not, our brains seem to be highly deterministic at the level that counts, at the level of the neural cell. So, against emotional sway, let us follow the evidence of science wherever it leads, independent of whether or not we find the implications emotionally appealing. Only then can we possibly discover that the truth, as initially uncomfortable as it might *feel*, turns out to be a better system than the one we had originally hoped for. Let us fully accept the truth about determinism and its necessary prevention of free will. Only then will we be able to see that free will is completely incompatible with two very desirable human traits: (1) the intellectual notion of rationality, and (2) the moral notion of reliability.

As I argued extensively in *The Laughing Genes*, "… the extent to which a mind is capable of rational thought is limited by the extent to which that mind is deterministic." After all, a free will allows one to freely choose irrational options. If a free will is constrained to always choosing the most rational option, then it is not free at all. And, as I also previously argued: "The degree to which a mind is capable of reliable behavior is limited to the degree to which that mind is deterministic." A free will encourages one to whimsically choose unreliable options, such as getting drunk instead of going to work. If a free will is constrained by commitment, duty and obligation, then it is not free at all. So, if you don't mind being completely irrational and unreliable, then you may wish for a mind with free will. I'll take a deterministic mind any day.

All relevant scientific evidence clearly shows that our brains were genetically designed to automatically identify and choose whatever paths they believe will maximize happiness. But unlike hedonists, who seek immediate gratification, our brains mostly pursue greater *future* happiness that is likely to result from the present investment of hard work. Thus, we tend to maximize the total happiness expected to be acquired over our lifetimes.

Happiness is not what we typically think it is. It does not benefit us in the manner that we believe it does. Our sources of happiness are defined for us by our genes as reward for our doing things that tend to perpetuate them – our genes. We are rewarded with feelings of pleasure when doing such things as having sex, eating food, seeking a comfortable environment, and raising healthy children. It is not a coincidence that our sources of pleasure tend to inspire activities that tend to benefit the perpetuation interests of our genes.

We have thus identified another critical premise underlying this book: All cognitive abilities, human and otherwise, are involuntary and automatic. Minds are brains, and brains are machines. The human ability to think exists only because it has statistically provided an evolutionary advantage to those replicating genes that have enabled the thinking. Even if free will were possible in a mostly deterministic universe, such a style of unpredictable thinking would be less desirable from evolution's perspective than a deterministic style of thinking that always seeks the best way to perpetuate whatever genes enable that thinking.

Nature selects capabilities based on their statistical benefits. And, even though our human minds sometimes make bad decisions on the basis of improper beliefs, they *always try* to maximize total future happiness based on the beliefs they hold. Some minds even choose death as a means to future happiness in heaven. Indeed, even such irrational events as terrorist suicide missions are inspired by expectations, however improper, of future reward. We'll later see quite clearly how our genes and memes interact to motivate our human behaviors in ways that *statistically* benefit the perpetuation of both the genes and the memes.

Given that human minds are mostly deterministic, it is perfectly logical to speculate that there may be some functional similarity between mental processes that automatically think creative thoughts and evolutionary processes that automatically evolve creative biological animals. Both happen through natural iteration of patterns by the strict laws of physics. But recognize that this perspective has dramatic implications regarding what we typically think of as 'intent'. Intent can never involve volition or free will. Intent is always involuntary and automatic. Even the intentions of humans must be involuntary, if indeed human brains are deterministic.

Intent must always result, automatically, from deterministic processes. Upon having accepted this new definition for 'intent', we needn't be shy in saying that all replicators, including memes and genes, actually do *intend* to seek ubiquity. We must simply keep in mind that *all* forms of intent are merely mechanistic, automatic and lacking in volition. Intent is nothing more than a predisposition to act toward a certain goal. This odd perspective allows us to say with impunity, for example, that a cocked and baited mousetrap fully *intends* to catch a mouse. And, consequently, I am forced to argue that the *feelings* of intent we humans experience amount to nothing more than illusions foisted upon us by our genes. Perhaps this will become more clear during later discussions of consciousness.

So, what is the essence of creativity? Is it the same for both the process of thinking and the process of evolving? Indeed, I believe it is. While we may identify some superficial differences, at bottom, creative thinking is just a process of evolution within a mind. As we'll later see, a mind can be any collection of material things operating in some coordinated fashion, cooperating toward the manipulation of patterns. A successful mind is one that acts so as to perpetuate the patterns that define it. For humans, those defining patterns involve cultural and technological memes just as much as genes.

Here is a critical point that will become ever more clear as the book proceeds: The process of human thinking involves simulation of the real-world environment. Within that simulated environment, the human mind can then consider different ideas by simulating their effects. But realize that since the real-world environment to be simulated is deterministic, the mental simulation of that environment must also be deterministic. That is, our minds absolutely must be deterministic in order to simulate, within our imaginations, the deterministic environment.

The mind of a human operates by manipulating patterns of thought – imagined behaviors – whereas the mind of biological evolution operates by manipulating patterns of DNA. When a human mind contemplates a problem, it imagines the problem paradigm over and over again in many various ways until it discovers a solution, just as evolution tries mutation after mutation to find a species that fills an ecological niche. It is the human imagination that is the simulator in which ideas can evolve.

As it turns out, the mind of man is not so different from the mind of God, so long as we think of God, our creator, as the process of evolution. We may rightfully say that God created us in his image. Indeed, evolution created our minds to operate in the image of evolution. If it appears as though I am trying to establish a common ground on which evolutionists and religionists may come together, realize it is not my primary intent. However, such a meeting of minds would not be a bad thing. It could lead to cooperative efforts that might mutually benefit everyone.

I am constantly amazed by elitist philosophers and ethologists who continually pound on the wedge that separates their political views from the views of others who think differently. As I previously argued in *The Laughing Genes*, the essence of morality is based on the notion of cooperation. Likewise, we'll find that the essence of intelligence also relies on cooperation, among neurons and among people. In fact, natural selection prefers both the characteristics of intelligence and morality as means for facilitating ever greater cooperation, So, the intelligent and moral thing for philosophers and ethologists to do is to remove the wedge and look for ways to broaden whatever common ground exists. I briefly allude to this concept as a 'teaser' meant to assure you that the pursuit of a deterministically automated intelligence does not require us to abandon our dearly-held ethical principles. In fact, it helps us to define them more clearly.

Now, let us return to exploring such mundane things as deterministically evolving patterns so that we may later properly appreciate the unification of morality and intelligence. If the essence of creative thought is the evolutionary manipulation of patterns, then, before we can hope to understand creative thought, we must understand exactly what it is that makes a pattern capable of evolving.

The Origins of Order

Patterns are key to understanding evolution and intelligence. So, what, exactly, is a pattern? Well, a pattern is nothing more than a *relationship* between multiple objects or events. When multiple things are related in space or time, they form a pattern. For instance, molecular patterns and crystal patterns exist as spatial arrangements of atoms – they are related in space. On the other hand, electrical pulses transmitted through a digital modem line form a much more temporal sort of pattern or relationship. The most interesting patterns are the ones that relate things in both space *and* time. For instance, the act of walking is characterized by a spatial relationship (between legs) that changes in a regular manner over time.

Many things in nature automatically form patterns. For instance, at the large end of scale, there tends to be a spiral pattern in the spatial relationship between the stars of a galaxy. At the small end of scale, certain types of atoms tend to automatically arrange themselves into highly regular crystal patterns under the right conditions. Atomic forces of attraction (and repulsion) are responsible for the stable spatial patterns of all sorts of molecules, including crystals.

As I think about patterns and the concept of order, I become intrigued by the idea that everything in the universe is arranged into some sort of ordered pattern. Absolutely *everything*! For instance, all the trees in a particular forest are arranged in a particular pattern. But that exact pattern is really quite meaningless. I can't even use it to identify other areas of the world that might qualify as forests, because no other forest in the world has the exact same pattern of trees.

The same is true for patterns of galaxies. Even though many galaxies take the shape of a spiral, we don't suspect that any two galaxies have identically positioned stars in identically shaped spirals. So, how can we say that many galaxies have the same pattern when in fact none are identical? How can we separate patterns into classes? What is the metric of similarity between patterns? These questions address *abstract* characteristics of relationships within patterns, which won't become clear until we examine the neural architecture of the human brain. For now, we must stay focused on the most fundamental characteristics of patterns.

To appreciate the natural creation of order at its most fundamental level, let us contemplate the beautiful patterns of snowflakes, which are composed of highly ordered patterns of snow crystals. The six-sided symmetry of snow crystals intrigued even astronomer Johannes Kepler, who, in 1611, turned from looking at the heavens to looking at snowflakes. In an essay titled *"On the Six-Cornered Snowflake,"* Kepler postulated that the symmetry of snow crystals arises from the geometrical efficiencies associated with well-packed arrays of tiny spheres. Three centuries later, scientists have come to only a slightly more precise understanding.

The hexagonal shape of a snow crystal is determined at the molecular level by the spatial arrangement of atoms composing the water molecule. Each water molecule contains two atoms of hydrogen joined to one atom of oxygen – H_2O – which naturally takes a 'Y' shape. Upon freezing, molecules of water can 'join hands' to form angles corresponding to those of a hexagon, thus providing the underlying shape and symmetry of the snow crystal lattice.

The complex diversity between various snow crystals results from the manner in which water molecules, floating in the air, attach themselves onto falling snow crystals. And that manner of attachment depends on local atmospheric conditions. Whether a nearby water molecule is more likely to attach itself to a corner of a growing hexagon or to an edge depends on local characteristics of temperature, pressure and humidity. Colder temperatures result in sharper ice crystal tips. Warmer temperatures cause ice crystals to grow more slowly and smoothly, resulting in less-intricate shapes. Higher pressures cause more intricate branching.

As a snow crystal blows about in a cloud, it experiences changing environmental characteristics of temperature, pressure and humidity. Thus, the growth characteristics of the six arms of a snow crystal change with time, reflecting the ever-changing conditions in the crystal's path. But because each arm sees roughly the same environmental conditions, each arm grows roughly the same way. James Gleick says it well in his book *Chaos* (1987, p.311): "The six tips of a snowflake, spreading within a millimeter space, feel the same temperature, and because the laws of growth are purely deterministic, they maintain a near perfect symmetry." Since no two snowflakes take the same path in their descent, it is highly unlikely that any two snowflakes will ever be alike.

In the late 1800's Wilson A. Bentley, a self-educated farmer living in Vermont, combined a microscope with a bellows camera to become the first person to photograph a single snow crystal. Over the course of more than forty years, "*Snowflake*" Bentley, as he became known, captured more than 5000 snow crystals on film. Bentley once said that "every crystal was a masterpiece of design and no one design was ever repeated. When a snowflake melted, that design was forever lost."

Should we care about the diversity of a snowflake design? It is surely tempting to draw an analogy between the value of unique diversity among snowflakes and the value of unique diversity among humans. But such a metaphor would indeed be misleading, for there is a very important difference between the intricacies of snowflake design and human design: Unlike the patterns of human genes, the pattern of each and every snowflake is designed from scratch, and is therefore extremely unlikely to ever serve as a template for its own replication or translation (unless it happens to get itself photographed).

Self-Replicating Patterns

Naturally occurring patterns cross an important threshold when they are able to cause the production of new instances of themselves. They then serve as templates for their own replication or translation. Such a capability lays the foundation for the emergence of life, but I'll later define another threshold, even more advanced than mere self-replication, that separates living patterns from non-living patterns. So, while life depends on patterns that get replicated, let us not yet think of replicating patterns as necessarily being instances of life.

There are indeed some naturally occurring patterns that automatically replicate themselves. For example, I've already mentioned that crystals form identical layers of lattice patterns as they grow. As new atoms attach themselves to the surface of a crystal, they tend to arrange themselves in exactly the same pattern as atoms already on the surface, to which they attach. So, layer by layer, patterns of atoms replicate themselves in newly forming layers. I'll later present some other instances of naturally-occurring self-replicating patterns. But for now, I want to focus on the fundamental principles of science that govern the creation and degradation of order. So, let us briefly consider the branch of science concerned with order – the thermodynamic concept of *entropy*.

It is a well-known scientific fact that entropy – a measure of randomness or disorder – always increases. That fact portends eventual doom for our universe, as it implies that matter is destined to become ever more homogenized, and energy is destined to become ever more uniformly distributed.

Even though universal entropy always increases, which means that the total order in the universe always decreases, local pockets of order can spontaneously emerge. And, even though highly improbable, an instance of a special kind of order – the auto-replicating kind – can also spontaneously emerge. And when it does, it dramatically creates more and more *local* order, but always at the expense of decreasing *universal* order.

It is a thermodynamic fact that an act of replicating any type of order necessarily requires the entropic dispersion – what we might call consumption – of available energy. So, replicators can become much more successful at reproducing themselves if they can somehow direct energy toward their own process of replication. Energy thus becomes a primary resource for replicating order. It is the fundamental reason why all instances of advanced life spend so much time chasing, gathering and eating food, and why plants have the means to photosynthesize sunlight into a chemical form of usable energy. It is also why the United States is so heavily embroiled in mid-East affairs: fundamentally, we need their oil to facilitate the ongoing evolution of our genes and memes.

With this very brief background into the thermodynamic nature of life, we are in a position to establish some fundamental thermodynamic principles that govern the natural evolution of patterns (similar to what I earlier presented in *The Laughing Genes*).

- The act of replicating an independent pattern always requires thermodynamic work and consumes available energy.

- Patterns that are able to direct energy toward their own replication tend to become increasingly plentiful.

- Mutated descendants of replicating patterns will occasionally produce patterns that are more efficient than their ancestors in their abilities to direct energy toward their own replication.

- Surviving descendants of mutating replicators, existing in a differentially selective and hostile environment, tend to become more efficient, on average, than their ancestors. But, in a benign and non-selective environment, the average efficiency of descendent generations tends to decrease simply because there are so many more inefficient patterns than there are efficient patterns.

- Patterns that mutually cooperate toward each other's abilities to replicate can be treated as a single pattern to an extent determined by the degree of inter-dependency (this defines a very important principle I previously termed the *inclusive phenotype*).

The last principle, regarding cooperation among patterns, is key to understanding morality. Additionally, the moral purpose of life is key to understanding nature's preference for intelligent life. Stated very simply, it takes intelligence to effectively cooperate. We'll explore the critical concept of cooperating patterns a little later.

When Order Becomes Information

We are now equipped to define an important threshold between order and information. As it turns out, the thermodynamic principle of entropy is intimately associated with the definition of information. In fact, the founder of *information theory*, Claude Shannon, derived a formula for expressing the information content of a signal based on its statistical improbability. That formula is functionally identical to the formula for calculating entropy.

Shannon's characterization of information has to do with *unpredictability*. We can analogize it in this manner: Suppose a friend of yours, a stock broker, gives you a stock tip, and all his stock tips tend to be very good ones. He has indeed expressed information to you. But suppose later in the day he gives you the very same stock tip again. Has he expressed more information to you? No, you might have expected the stock tip he was to later give. It was predictable, and therefore, devoid of information.

Only when ordered data is unexpected or unpredictable can it be deemed as information. But this leads to the unsatisfying realization that the type of signal expressing the most information (according to Shannon) is a completely random signal. From an information theorist's point of view, this is true. A random signal is as unpredictable as you can get, and is therefore completely unable to be compressed. The unpredictability of a random signal causes it to require more bandwidth for transmission than a similar signal having predictable patterns of compressible regularities. Indeed, Shannon's entropic definition for information correlates perfectly with the required bandwidth for transmission, but it is quite contrary to how we normally think of information. If your stock broker friend mumbles random sounds in your ear, Shannon will declare that he has thereby conveyed more information to you than if he had, for example, told you to 'buy Google'.

Whereas Shannon's statistical definition for information is valuable from an engineering standpoint, it is not consistent with our intuitions. Nor is it useful, philosophically. The way we typically think of information depends on something much more difficult to quantify. It depends on the ability to *benefit* from the particular order involved. Whereas everything in the world is ordered in some particular way, only a few types of order provide benefit, and thereby represent information. For example, consider how a monkey banging on a typewriter arranges letters in a particular order on a page. Shannon's formula for calculating information content may in fact assess a high level of information (unpredictability) in the random arrangement of letters, but most humans would agree that the particular arrangement of letters is only information if it can be read, understood and used for benefit.

Fundamentally, the order of a pattern only has *value*, and thereby becomes information, if it can be used by some other pattern to perpetuate its own particular order. Returning to the previous example, if your broker friend gives you stock tips that always lose money, then you would not consider his data to have any value. And, without value, it cannot be information. Surprisingly, the threshold between order and information is dependent on the slippery characteristic of observer comprehension and benefit.

We have uncovered an interesting philosophical issue. How can we say that a pattern has value? That assertion would seem to contradict philosopher David Hume's conjecture that there can be no sort of implication drawn from *is* to *ought*, from *fact* to *value*. There is no equivalence or conversion factor between the dimensional units of *what is*, and *what ought to be*. But indeed, patterns do have value if they are able to facilitate the ongoing replication of order. Despite what most philosophers might believe, the perpetuation of order is necessarily at the root of all *value*. For humans, there is always a causal pathway connecting anything of *importance* to human survival or procreation. Indeed, survival and procreation represent the perpetuation of order critical to the human species.

Here is a clinching argument to support the idea that order only becomes information when it benefits the furthering of life. If we were to perform a particular operation on a newly-fertilized human egg cell, by scrambling a randomly selected 10-nucleotide stretch of one of its chromosomes, there would likely be no effect on the ontogenetic development of the consequent human. Why? Because less than five percent of a human's DNA is used to describe the construction of its physical body. Only that very small portion of its DNA codes for proteins. The rest is junk filler that has hitched a ride using the existing mechanism for copying the useful DNA. But notice that, when we refer to most of a human's chromosomes as junk, we are making a value judgment. We have every right to make that valuation, simply because we have every right to think our human lives have value.

So, order only has meaning or value, and thereby becomes information, when it provides benefit to life. For example, the arrangement of order in some arbitrarily complex molecule is meaningless and void of information unless it represents an ingredient for some aspect of some sort of life. DNA, on the other hand, is full of value and meaning because it creates life. But in a human's chromosomes, only the small fraction of DNA that codes for life is informational.

Thinking back to the earlier example, regarding the ordered pattern of trees in a forest, the particular arrangement is meaningless unless it is copied onto a piece of paper as a map to be used for navigation. The order of the trees both in the forest and on the map – having jumped from one substrate to another – now becomes information to someone who can use it for benefit. I find it a bit odd that it takes life to assess such things as information content, meaning and value. But the world gets even stranger when we descend to sub-atomic scales – to the quantum level.

Just as order needs life to assess its value and its informational content, so also is there a similar sort of dependence on the need for life to arbitrate fundamental reality at the quantum level of physics. Hold onto your hat, because this gets mind-blowingly weird. Physicists have understood for quite some time that both a particle's position and momentum can never be simultaneously known with certainty. This is due to the 'fuzzy' nature of sub-atomic particles, to which I earlier referred, known as Heisenberg's uncertainty principle. While some scientists have characterized the inherent uncertainty of nature as a problem of measuring, it is now believed by most physicists that the problem may actually be a much more fundamental one. It is not just a matter that we can't precisely measure a sub-atomic particle; experiments show that even nature herself is uncertain of the exact position and momentum of any given particle.

At any moment, a given particle exists only as a spatial probability distribution. In some sense, scientists argue, the particle exists at all possible locations within that distribution, all at the same time. But as a particle interacts with another particle, nature is forced to make a definite decision as to where both particles actually are. After all, the resulting paths of two particles, just after they collide, depend entirely on exactly where they were and where they were headed just before the collision. So, nature must make up her mind at that point. Or must she? Nature could just allow the uncertainty to accumulate. In fact, a popular interpretation of quantum dynamics does indeed suggest that the uncertainty of all quantum events accumulates *until a sentient observation is made*. At that moment, the many chains of interaction possibilities that could have happened collapse down to a single chain of dependent interactions that actually did happen. In essence, nature makes up her mind.

This weird situation, regarding the need for observer participancy to force nature's hand, is well known among physicists as the *Copenhagen interpretation* of quantum dynamics. It is described mathematically by Erwin Schrödinger's wave equation. Some very smart scientists discovered it (Niels Bohr and Werner Heisenberg), and Einstein wrestled with it until his death. I'll later speculate on an interesting cosmological reconciliation of this odd phenomenon of nature with the meaning and purpose of intelligence. But, I warn you, it will take an open mind to appreciate it.

Is there a connection between the way that order becomes information only when it benefits life and the way that nature resolves quantum uncertainty only when life participates? Perhaps not. But we should find their odd similarities a bit interesting. It is probably the case that we humans have simply *defined* 'information' as only the subset of order that provides benefit to us. When we realize this implicit definition, the mystery begins to disappear. However, the explanation I'll offer regarding the mystery of quantum uncertainty will indeed cause us to suspect that life is the focus and purpose of the universe. That anthropocentric explanation will have to wait for the development of many other concepts.

So, what does this all have to do with memes? Well, the concept of memes tends to involve patterns that benefit life in some way. Indeed, aside from the patterns of DNA that create life, the most important patterns of order are the ones produced *by* life for the benefit *of* life. For humans, those are primarily memetic patterns of behavior – the things we do to stay alive and raise our children, and the things we do to produce useful artifacts. We'll eventually find that all beneficial behavioral patterns tend to convey information to other instances of life that are capable of mimicking those behaviors.

We'll soon see why natural selection preferred genes that enabled humans to mimic the behavioral patterns of others. Indeed, we'll see why evolution found mimicry of some behaviors to be extremely valuable. And we'll see how that human propensity for mimicry, from person to person, enabled an explosive mechanism for massively replicating patterns of behavior. Some commonly replicated behaviors are informational in the way that they support the furthering of life, but many are not. Many cultural behaviors, such as fads and rituals, are like the stretches of junk DNA in the human genome that have hitched a ride using the very same copying mechanism responsible for replicating the useful stretches of DNA. And so, we'll also see how some non-beneficial patterns of behavior can be automatically propagated by the same mechanism of mimicry that was genetically designed for propagating useful behaviors from parents to children.

Before we tackle those interesting issues, we need to trudge through some more fundamental concepts. We've come a long way toward understanding what patterns are and how they can become more complex through natural processes, but we haven't yet officially designated at what point the deterministic process of evolving complexity becomes what we think of as *life*. So, that will be the focus of the next chapter.

Chapter 2. **The Fundamentals of Life**

In the ongoing debate over how life might have gotten started, doubters of evolution cite the extremely low probability of atoms randomly coming together to form the original complex molecules of DNA. Making the situation even more improbable, there also had to have been the simultaneous assemblage of the necessary associated molecular machinery for processing the DNA. The typical and somewhat unsatisfying answer to this challenge relies on a great expanse of time and a huge number of random combinations of atoms.

The argument is similar to the one that, given enough time and enough monkeys sitting at typewriters, one of them will eventually write something interesting. But monkeys do not type randomly, and by analogy, atoms do not assemble themselves randomly. Just as monkeys may favor some keys over others, or may tend to press the same key repeatedly, atoms floating freely in solution have clear preferences for linking up with certain complementary atoms.

Perhaps the natural affinities between certain combinations of atoms were designed by an intelligent creator of our universe so that self-replicating molecules, such as DNA, would be more likely to randomly assemble than other sorts of molecules of similar complexity. I'll later present some logical evidence indicating that the universe and the forces of nature may have been carefully tuned, and were therefore probabilistically destined to create something like humans. The evidence suggests that the intelligent tuning is likely to have emerged, not from an anthropocentric sort of god, but from an evolutionary process operating at the level of the Cosmos on universal constants. But, for now, let us continue building a strong intellectual foundation by speculating on how life might have gotten started.

Crystal Patterns and Autocatalytic Sets

An interesting approach to the problem of 'bootstrapping' the evolutionary process is suggested by Graham Cairns-Smith in his book *Seven Clues to the Origin of Life* (1985). The essence of his argument speculates on an evolutionary process that can occur as crystals grow. I'll briefly summarize the idea.

Diamonds are able to be cut into beautifully symmetrical shapes because of the precise arrangement of their atoms. All pure and perfect crystals have their atoms precisely ordered into a highly regular lattice structure. There are many types of crystals that occur naturally, and scientists have learned a great deal about the conditions under which they can develop. Crystals can be grown in the laboratory by allowing a molten solution, saturated with crystallizable atoms, to cool in a slow and controlled manner. As atoms in the solution lose their energy they will tend to attach themselves to the surface of an existing crystallized mass, and they will do so in a manner that repeats the organized pattern of alignment in the atoms of the existing mass. The process generally requires a 'seed' crystal whose atoms are already aligned, to which the condensing atoms can attach themselves. As each layer of atoms becomes attached to the face of a crystal, the alignment pattern of the preceding layer is thereby replicated. Recall that replication of a pattern is a key ingredient in the process of evolution.

Occasionally, there can be a defect in the alignment pattern of a crystal – a point on the surface where an atom is missing, or where a contaminant atom of a slightly different type has been randomly incorporated. The atoms surrounding a defect will adjust their positions, taking on a particular new pattern that is unique to the style of that particular defect. As a visual analogy, we might imagine a fabric with a slight 'pull' in one of the threads. Alignment defect patterns can then be roughly replicated by new layers growing on top of the defect layer, as the new layers become attached. We have thus formed the basis for mutations in the crystal growing process. The effects of multiple defects can accumulate as layers are added, and the accumulated result can be any one of many unique replicating patterns.

Without going into significant detail here, Cairns-Smith speculates on an environmental selection process in naturally occurring clay mud that favors some crystal-defect growth patterns over others, and a means for distributing those patterns geographically as seeds for new crystal growth. In short, crystals grow to a size large enough that parts of the crystals break off, forming new crystals with similar or identical patterns of growth. If the seed pieces are carried by trickles of subterranean water flows to new destinations, then diverse types of order can freely spread across wide geographic areas. Many diverse crystal patterns could have naturally evolved from this simple mechanism. These diverse patterns could have acted in the manner of a kind of scaffolding for the formation of many complex and unique types of molecules. The patterns could have provided varying sorts of attraction sites to hold various types of atoms in positions while bonding, thereby enabling the formation of many sorts of molecules, including perhaps even crude DNA structures.

While I find it difficult to make the mental leap from complex crystal patterns to complex DNA molecules, that scenario is easier to embrace than the idea that DNA molecules would have come together by pure chance. Going back to our analogy of monkeys at typewriters, they would certainly have a better chance at getting published if they were allowed to use stenograph[2] machines, instead of typewriters. The higher level of coding in a stenograph machine captures more complexity, and therefore gets the monkeys closer to the goal of getting published with each and every keystroke. In a similar manner, the complexity resulting from crystal growth could have gotten molecular elements closer to the goal of self replication.

The notion that abstract complexity evolves in some manner, whether it be through crystal growth or some other repetitive process, is scientifically parsimonious and therefore satisfying. By conceiving of a process that simply cranks out lots of different complex molecules, we elevate the unlikely assemblage of DNA to a statistical inevitability.

Cairns-Smith's idea illustrates an important sort of phenomenon: It is logical to suspect that abstract complexity may indeed occasionally jump from one substrate to another. According to Cairns-Smith, it may have jumped from corrupted crystal patterns to molecular patterns. This concept will become very important in our discussion of memes, as the complexity of our human genes now seems to be fostering a corresponding sort of complexity in technology. Complexity itself may be in the transitional phase of jumping from one substrate to another, perhaps for the second or third time in the history of Earth.

Another method of 'bootstrapping' the process of evolving complexity comes in the form of what I'll call circularly self-catalyzing sets of molecules. In the domain of chemistry, a catalyst is the presence of a particular substance that aids in the combining of other substances.

Perhaps we can conceptualize a model of evolving complexity using familiar chemical catalysis. Here is how Stuart Kauffman, of the Santa Fe Institute, thought about it:

> ... suppose, thought Kauffman, just suppose that some of these smallish molecules floating around in the primordial soup were able to act as "catalysts" – submicroscopic matchmakers. Chemists see this sort of thing all the time: one molecule, the catalyst, grabs two other molecules as they go tumbling by and brings them together, so that they can interact and fuse very quickly. Then the catalyst releases the newly wedded pair, grabs another pair, and so on. ... All right, thought Kauffman, imagine that you had a primordial soup containing some molecule A that was busily catalyzing the formation of another molecule B. ... Now, thought Kauffman, suppose that molecule B itself had a weak catalytic effect so that it boosted the production of some molecule C. And suppose that C also acted as a catalyst, and so on. If the pot of primordial soup was big enough, he reasoned, and if there were enough different kinds of molecules in there to start with, then somewhere down the line you might very well have found a molecule Z that closed the loop and catalyzed the creation of A. ... The compounds in the soup could have a coherent, self-reinforcing web of reactions. ... Taken as a whole, in short, the web would have catalyzed its own formation. It would have been an "autocatalytic set." ... it meant that life could indeed have bootstrapped itself into existence from very simple molecules. And it meant that life had not been just a random accident, but was part of nature's incessant compulsion for self-organization.

> **– M. Mitchell Waldrop,** *Complexity* (1991, pp.123-124)

That last sentence, regarding "nature's incessant compulsion for self-organization," is a theme I find to be profoundly meaningful. You'll understand the significance if you've ever seen a dynamic presentation of a fractal that zooms in, ever deeper, into the infinitely deep and uniquely beautiful patterns that mathematically exist at any region along its edge. The beauty was already there, long before life ever existed, just waiting to be discovered. The universe is like a constantly unfolding fractal, slowly revealing its evolving patterns of never-ending complexity. The delicate balance between order and chaos, between uniformity and variety, between beauty and sterile utility, is partially unveiled during the gradual blossoming of a thorny rose. The profound implications of nature's obvious compulsion will pervade the rare philosophical moments of this book.

Kauffman's description of autocatalytic sets takes on even more credibility when we realize that the modern process of natural biological reproduction is best described as a circular, self-catalyzing reaction. Briefly stated, enzymes operate on DNA molecules, by a process known as transcription, to assemble amino acids into proteins, some of which are themselves the very same enzymes. Each stage of the circular process is merely a chemical reaction that is catalyzed by the products of the previous stage. Perhaps the principle of catalysis is responsible for having 'bootstrapped' the process of biological evolution. Indeed, it is now believed by many molecular biologists that a form of RNA could have been just such a self-catalyzing type of molecule that might have come to exist in the primordial ooze. The role of RNA in modern biology is to act as an intermediate product in the process of transcribing DNA to proteins.

The important point here is that there are many avenues by which various patterns of ordered complexity are able to evolve toward greater complexity over time. It is not crucial to know exactly which one led to the first instance of DNA. It is only important that we find some mechanism by which such complexity could have naturally emerged.

Catalysis is a powerful concept. Even the evolution of our modern economic system depends on it. Tools, machines, trucks, and factories – all sorts of capital equipment – *catalyze* the bringing together of raw materials and human labor in the production of all goods and services. And some of those goods are themselves more modern types of capital equipment – such as computers – that catalyze a whole new level of better goods, services and yet even more modern forms of capital equipment. Catalysis is a form of cooperation that automatically occurs between a catalyst and two primitive ingredients. The automatic process yields a product that didn't exist before, and as a result, has the potential for creating synergy out of nothingness. Synergistic cooperation, in many different forms, is the essence of life, and life is the eventual product of self-replicating order.

The emergence of order from certain dynamical systems seems to be built right into the laws of nature. We can learn something about the fundamental origins of order through the science of complexity theory, which is the study of how complex order naturally emerges from many types of nonlinear systems. Amazingly, all dynamical systems that descend into chaotic behaviors tend to exhibit the same distinctive sorts of patterns as they transition between order and turbulent chaos. Mitchell Feigenbaum, a scientist at Los Alamos National Laboratory, discovered that there is a region between order and chaos in which regular events happen, and they happen at a regular rate:

> With his calculator he began to use a combination of analytic algebra and numerical exploration to piece together an understanding of the quadratic map, concentrating on the boundary region between order and chaos. ... En route to chaos in this region was a cascade of period-doublings, the splitting of two-cycles into four-cycles, four-cycles into eight-cycles, and so on. ... The period doublings were not just coming faster and faster, but they were coming faster and faster at a constant rate. ... Feigenbaum calculated the ratio of convergence to the finest precision possible on his machine – three decimal places – and came up with a number, 4.669.
>
> **–James Gleick**, *Chaos* (1987, pp.171-172)

Feigenbaum went on to calculate the ratio of convergence for a different type of nonlinear mathematical function and found that he got the same number.

> Incredibly, this trigonometric function was not just displaying a consistent, geometric regularity. It was displaying a regularity that was numerically *identical* to that of a much simpler function. No mathematical or physical theory existed to explain why two equations so different in form and meaning should lead to the same result.
>
> **–James Gleick**, *Chaos* (1987, p.173)

Complexity theory shows that a mathematically predictable mixture of complexity and order will naturally emerge from certain nonlinear systems, so long as they are fed with energy and poised on the edge of chaos. We'll find that being poised on the edge of chaos is a recurrent theme in the successful evolution of all replicators, and that certain nonlinearities in the human brain do indeed balance the activity of neurons on the edge of chaos.

Within this context of complexity theory, it seems that the eventual emergence of a self-replicating set of molecules was statistically destined to occur, given enough time and energy. The statistical destiny, regarding the emergence of self-replicating molecules, was written right into the laws of nature. But at what point do self-replicating sets of molecules become life?

What is Life?

We may define the word 'life' any way we want to. And, just as we may declare a fetus to be 'alive' at any of several thresholds – the onset of a heart beat, the achievement of a certain mass or age, or the moment of birth – we may similarly choose among several definitions for the threshold at which the phylogeny of life begins. We may say that the mere act of self-replication begins the process of life, but I prefer a more advanced threshold. I define life to start at the moment when self-replicating molecules are able to *direct energy toward their own replication*. At that point, the molecules actually have an active mission, however determined it may be. The mission of gathering and using energy gives the replicating patterns a purpose. After that critical threshold, the patterns have a sort of *intent* to perpetuate themselves, albeit involuntary (recall that, if the human mind is deterministic, then even human intent must be automatic and thus involuntary). Intent and purpose are certainly characteristics of what we typically think of as life.

The performance of any physical action requires energy, so the mere splitting apart of the two complementary halves of a DNA strand certainly requires some energy. And for that reason, life evolves so as to become ever more adept at gathering energy. Plants, for example, produce molecules of chlorophyll in order to gather energy from sunlight. The molecules of chlorophyll absorb energetic photons and then use that energy to assemble carbohydrates, which are used to construct various parts of the plants. And, animal life is only possible due to the copious amounts of energy collected and stored by plant life. Animals get their required energy by eating plants, or by eating other animals that eat plants, or by eating animals that eat animals that eat plants. In most cases, animals ultimately get their energy from plants, which get their energy from sunlight.

We can't logically say that primitive self-replicating molecules gathered energy. Instead, they got their required energy from the Brownian motion of surrounding molecules. But there did come a point in the development of earthly life when patterns became capable of gathering and storing energy for perpetuating themselves. For instance, some bacteria 'grow' a flagellum when they sense a local depletion of ingredients that are necessary for replication. The flagellum is a wiggling tail that propels its bacterium to a new region where required ingredients may be more plentiful. Such propelling action clearly directs energy toward the perpetuation of its defining replicators. This threshold for defining when life begins establishes that self-replicating molecules lacking motility and drifting aimlessly in the ooze *are not* life, but bacteria that 'swim' to nutrient-rich destinations *are* life.

The collection and direction of energy toward the perpetuation of replicating patterns is crucial to all life. And the degree to which a species is evolved correlates well with its ability to direct energy toward the perpetuation of its own defining replicators. Indeed, humans have harnessed all sorts of energy in the forms of coal, petroleum, natural gas, various other chemical reactions, and even nuclear reactions. Because we humans are able to use energy so effectively, we have become the dominant species. In essence, we direct energy toward the perpetuation of our genes better than any other species does. Recognize that one way to direct energy toward the perpetuation of replicating patterns is by eliminating other competing replicating patterns. Thus, the forceful use of weapons against enemies is certainly a way of directing energy toward the replication of one's own defining patterns.

We have discovered a natural sort of purpose at the most fundamental level of life. The natural purpose of a replicator is to replicate. And, thus, a *good* replicator is one that is able to facilitate its own replication. Indeed, it must be the case that all *value* springs from whatever facilitates the perpetuation of purposeful replicators. For example, there is no way to discern among several crystal patterns which is 'better'; any crystal pattern is just as 'good' as any other. But things change dramatically when patterns cross the threshold of life, as I've described it. When patterns facilitate their own replication, then we may logically say that patterns replicating more prolifically are *better*. Perhaps we can make that claim even before patterns act for their own benefit, but the claim takes on real significance when patterns exhibit an apparent 'self-interest', by directing energy toward their own replication.

A pattern's ability to facilitate its own perpetuation may be the only fundamental definition for 'goodness' in the universe, however bland you may think it to be. We now have a perfect explanation for why we humans place such a high moral value on the health and well-being of all children, especially our own. We are programmed to do so by our genes. That is how they direct energy toward their own perpetuation. They have intelligently evolved to be that way.

Patterns of Behavior

If you've ever looked through a microscope at a cell undergoing replication, you'll know that it just sort of sits there, floating aimlessly in its environment, minding its own business, until its nucleus suddenly and spontaneously begins to divide. Of course, there is actually no spontaneity involved. The action occurs according to the strict laws of physics. The visual act of splitting apart only occurs when other invisible processes have reached some critical threshold. Those invisible processes involve the migration of ions through various membranes, and other sorts of events at the atomic level. Because the underlying processes are invisible to us, we are deceived into thinking that the visible events are spontaneous.

The same sort of deception happens at much larger scales when, for instance, we look at a child sleeping. It appears so calm and peacefully still; but inside the child a heart beats in reliable perpetuity, blood pulses through veins and arteries, neural firings cascade through complicated loops of brain activity, and within every cell, molecules are constantly engaged in an amazing dance of complex interactions. All this goes on continuously behind the closed eyes and the expressionless smile. Life is not really as we see it. Our conscious experiences deceive us into thinking we freely choose our behaviors on the basis of nothing more than whim and fancy. But a memetic perspective reveals the truth.

Human behaviors take on a whole different flavor when analyzed memetically. We typically analyze our own behaviors by looking at the ways we humans use them to serve our own purposes. But we need to think memetically by taking a meme's-eye-view of behaviors, just as Dawkins taught us to think genetically by taking a gene's-eye-view of our biology. The view of reality is always much clearer from the perspective of replicating patterns than from any other perspective. The odd truth is, patterns of memes exist for their own perpetuation benefit, not necessarily for our benefit.

In our deterministic universe, everything has a cause. Evidence shows that behaviors tend to be caused by replicators – either genes or memes – not by any sort of volitional free will.

Consider how genes evolve. Through generations of evolution, genes 'gather' knowledge about their environments, and they thereby become more intelligent in the ways they produce their host bodies. Genes have gotten so intelligent – so good at assembling efficient hosts – that in many species they produce hosts that are themselves intelligent. The hosts are able to learn from their respective environments and use that information to better perpetuate their genes. This allows them to adapt even to environments that are very different from the environments of their ancestors.

Genes automatically gather knowledge in their molecular organizations so as to build more intelligent hosts. And intelligent hosts, in addition to having some built-in knowledge, automatically gather even more knowledge in their neural organizations in ways that tend to perpetuate their defining genes. We'll later see exactly how that acquired knowledge comes in discrete packets, each of which corresponds to a meme. The gathering of knowledge, in either genes or memes, happens automatically through natural evolutionary processes. And that knowledge, then, automatically inspires intelligent behaviors. So, let us contemplate both sorts of replicating causes of behavior, starting with those inspired by genes.

Most animals are genetically designed to sense their environments in ways that allow them to detect the presence or absence of certain conditions. Various conditions are recognized by the existence or non-existence of associated patterns. When certain conditions are detected, then genetically prescribed activities are automatically executed. Just as a mousetrap snaps its bar when it detects the presence of a mouse, so do automatic mechanisms in living things execute certain genetically defined behavior patterns in response to the detection of certain conditions. I'll sometimes refer to these automatic mechanisms as *responders*.

Even the lowly bacteria that sprout flagella when they sense a local depletion of necessary ingredients are exhibiting intelligence in the act of deciding when to move to a different location. Their behavior is merely contingent upon a certain condition in the environment. That style of contingency programming is how intelligence first manifests itself in the phylogenetic development of life: 'If I sense *this*, then I do a certain activity; if I sense *that*, then I do a different activity'. Now that we have a fundamental understanding of how genes inspire life-enhancing behaviors, let us begin to consider the behaviors inspired by memes.

Life-enhancing patterns of actions, in addition to being genetically programmed as automatic responses to certain perceived characteristics of the environment, can also be mimicked from other instances of life. When patterns of actions – behaviors – are commonly mimicked from one instance of life to another, then there emerges a new style of replicator involving spatio-temporal patterns that serve as templates for their own replication. We have already defined those mimicked behaviors as 'memes'.

We ultimately want to understand why genes bother to build brains that automatically collect memes. What we'll find is that memes enable mimicry, mimicry enables culture, and culture allows humans to adapt to changing environments much more quickly than genes can adapt. But then the critical question is: How can genes ensure that memes will evolve in ways that benefit them – the genes? The answer is: they can't. Some will and some won't, but the ones that do will enable a synergy that benefits both styles of replicators – genes and memes. This is a crucial concept that we must address before doing anything else. We must understand how multiple patterns can sometimes work together toward each others' benefit, which must result in their mutual perpetuation.

Cooperating Patterns

Multiple instances of patterns can sometimes mutually benefit each others' perpetuation, and when they do, they may effectively be treated as a single pattern. Despite the simplicity of this concept, I believe it represents the fundamental essence of life. Yet, it requires a slightly new way of thinking. Let us consider a very simple example that is intended only to set the concept in your mind.

Imagine a 2x2 grid of squares arranged so that the upper left and lower right squares are white and the upper right and lower left squares are black. Let us refer to this, the fundamental building block of a checkerboard, as pattern A. By arranging four A's into a square, we get a slightly larger 4x4 checkerboard – we'll call it pattern B. By arranging four B's into a square, we thereby assemble a full 8x8 checkerboard, finally suitable for playing chess or checkers. Let us think of this new pattern, C, as having achieved some sort of synergy by combining multiple constituent patterns. We can't play chess or checkers with either pattern A or pattern B, but we can with pattern C.

We may logically consider multiple instances of the smaller patterns as being able to *cooperate* with each other so as to create the bigger pattern. Now, of course, when I speak of patterns cooperating, I am not implying that they make some sort of a deal with each other. I simply mean that they just happen to mutually enhance each other's chances of getting perpetuated. Certainly, the many 2x2 grids of squares that form millions of checkerboards all over the world have enjoyed great prosperity as a result of their having been cooperatively combined into 8x8 grids of squares. Now, I do realize this very simple example is somewhat awkward because the small patterns are cooperating with other instances of themselves. But we'll soon see many cases where the principle of cooperation applies among patterns that are vastly different, sometimes even existing in different substrates.

Patterns of genes represent the prototypical example of cooperating patterns, as they clearly exhibit a very powerful cooperative synergy. Of the thousands of genes in the human genome, none are capable of perpetuating themselves without some of the others. Taken together, they are capable of constructing the fittest form of life on Earth. They most definitely cooperate toward their mutual perpetuation.

Cooperating patterns need not necessarily be proximally connected in order to generate synergy. Consider, for example, the genes of flowers and the genes of bees. They certainly cooperate toward their mutual perpetuation. Flowers have genes that cause them to produce nectar, which benefits the genes of bees. Bees have genes that cause them to carry pollen from flower to flower, which benefits the genes of flowers. The genes of both, flowers and bees, critically depend on each other for their mutual perpetuation, even though they reside in completely different species, indeed in completely different kingdoms. Just as two halves of a checkered game-board combine to become a single useful pattern, so may the genes of bees and the genes of flowers be considered as something of a single evolving pattern.

In the natural course of evolution, replicating patterns sometimes cause effects that, merely by chance, happen to facilitate the perpetuation of other replicating patterns. When two different patterns cause actions that mutually benefit each other, then both of those patterns perpetuate more readily as a result of the cooperation. It is the adaptive benefit of cooperation among replicators and among the effects they produce that represents the very essence of life and gives us humans our morality.

Morality is concerned with how instances of life should interact with other instances of life. Nature's preferred morality seems to be defined as follows: Life at any level acts morally when it achieves symbiosis, forms alliances, or otherwise acts cooperatively toward mutual benefit with whomever or whatever it interacts.

The synergy resulting from cooperation gives the cooperating participants an evolutionary advantage over other similar but non-cooperating instances of life. We see the synergistic effect at the molecular level in the form of cooperating genes that construct the various cellular mechanisms needed for protection and replication. We see the effect at the level of cooperating cells that work together to form cooperating hearts, lungs, arteries and veins. We see the effect at the individual level in the form of friends, families, colonies, packs, herds, flocks and hives. And we see the effect at the group level in the form of treaties and alliances.

In what is destined to be widely recognized as a landmark philosophical event, Robert Axelrod showed that the nature of human morality is likely based on the principle of cooperation. His book *The Evolution of Cooperation* (1984) reveals a cooperative style of morality as the optimal strategy for achieving inclusive fitness in any environment where there are others who are capable of cooperating. In the logically simulated worlds of game theory, the optimal strategy for all players in any multi-player game is to cooperate toward their mutual benefit, if indeed the game represents some aspect of real life.

In the now-famous game of Prisoner's Dilemma[3], the optimal strategy is known as *Tit-for-Tat*. The strategy, appropriately enough from a memetic perspective, is reminiscent of mimicry. By mimicking the other player's most recent action, an individual can discourage self-serving behaviors and encourage cooperative behaviors on the part of the other player. A strategy of mimicry by both players can lead to either a spiral of descent or a spiral of ascent, depending on the initial behavior.

If you help me, I will help you. But, if you betray me, I will betray you. Such is the spirit of mimicry that manifests itself in the human emotions of gratitude and revenge. Indeed, it doesn't take much intelligence to realize that reciprocated acts of gratitude are far preferable to reciprocated acts of vengeance. But it does take some intelligence to remember which of one's many associates tend to cooperate and which tend to betray. And, thus, one of the most important effects of intelligence is to encourage cooperative behaviors. Such is the basis of our human moral values. Indeed, we'll discover a great deal of evidence suggesting that the real evolutionary purpose of intelligence is to enable and enforce a cooperative style of morality.

Neighbors might conduct an ongoing feud, or they might enjoy a mutually beneficial friendship. The history of a relationship typically determines its future, as behavioral patterns from the past are perpetually and reciprocally mimicked into the future. As a general rule, since the long-term consequences of a relationship depend so heavily on initial behavior, it is wise to be nice when meeting new people for the first time. Indeed, we humans tend to be programmed in just such a manner. For instance, the act of shaking hands indicates at the outset an intent to be cooperative.

Sharing neighbors need only one set of tools between them. Cooperating neighbors can engage in productive activities – big construction projects – that require multiple people to perform, such as barn-raisings. And, caring neighbors help each other through difficult times. It should be obvious that people engaged in ascending spirals of reciprocated assistance will ultimately find themselves to be much more fit than those engaged in descending spirals of recrimination. So, the morality encouraged by nature is one based on mutually beneficial cooperation – I like to call it *cooperationism*. We may logically speculate that all instances of life, everywhere in the universe, will eventually evolve behaviors that are cooperative in nature. They will all develop moral values similar to ours. I'll later show that there is an arrow of directionality in the way that all life tends to evolve. There are gross stages through which the development of all life will eventually pass, and moral goodness appears to be something of a final destination stage (unless there are yet higher stages still to come).

By considering the synergy that can result from cooperation among multiple patterns, we can begin to lay the foundation for a moral framework that is consistent with nature's goals. Natural selection gives preference to a certain style of behavior that might initially appear as a selfish style of morality, but the character of what it means to be selfish will change as we come to appreciate what it is that actually defines the 'self'. The fundamental purpose for human morality is to cooperate toward the emergence of an individuality at a higher level – a higher order of life, and a bigger 'self'.

Cooperation between many individuals accumulates their strengths, and produces synergistic benefits of efficiency. Such cooperative unification is preferred by natural selection because it makes all the participants more likely to perpetuate their respective defining replicators. For instance, individuals are encouraged by nature to embrace a patriotic attitude toward their society (the larger self), by working cooperatively with their compatriots. Indeed, our genes imbue us with desires for joining, identifying with, and sacrificing for, various sorts of groups. It is why some of us enlist in the armed forces during wartime. It is why some of us emphatically root for local sports teams. It is why some of us are willing to go way beyond the call of duty for the company.

Cooperation produces synergy out of nothingness. It produces a system that is greater than the sum of its parts. We should have no trouble seeing why nature would prefer genes producing people that cooperate together so as to become more survivable and more efficient at replicating their genes than they otherwise would be. Indeed, the love between parents is more properly viewed as a genetic mechanism encouraging them to cooperate toward the raising of their children.

Metaphorically, our genes have good reason to laugh at how we humans unwittingly define our morality to serve their perpetuation interests – the interests of our genes – even above our own interests as individuals (see *The Laughing Genes*). Indeed, our genes coerce many of us, through our parental feelings, into sacrificing dearly for our children. Our various feelings of love cause most of us to get married, have babies and then work together to care for those babies. Even most childless people similarly feel that humanity's highest moral value is to protect innocent children. The benefits of all those feelings accrue to the perpetuation interests of our genes, not our individual selves; although, we are made to feel, by our genes, as though the benefit accrues to us as individuals.

We've seen how the genes of humans cooperate with each other toward the production of humans who are themselves cooperative. But, the benefits of cooperation are not limited to genes. They extend to all patterns that cooperate toward their mutual perpetuation. Indeed, they apply to patterns of memes as well. We don't usually think of inanimate objects as being capable of cooperating, but let us open our minds to such a view.

We see the synergistic effects of cooperation in the memetic domain of technology where, for example, the highly specific patterns of materials found in electrical components, such as transistors, resistors and capacitors, can combine and cooperate in various ways to produce very useful and often-replicated patterns of electronic devices. Various cooperating patterns etched on silicon integrated circuit chips have been replicated billions of times simply because of their enormous synergistic effects, especially in cooperation with humans as enhancements to cognition.

While we humans like to think of the situation the other way around – that we use silicon memory chips to improve our lives – from the perspective of replicating patterns, humans merely represent the 'nutrient-rich' environment in which memory chips are able to get their patterns reproduced. If human minds are indeed deterministic, then the sterile perspective from the vantage point of replicating patterns is the one that makes the most sense.

The effects of cooperation occur at many different levels. At any level, a sort of unification occurs as a result of cooperation among multiple patterns. And, multiple *sets of unified patterns* further cooperate toward *their* unification at an even higher level. The result is a hierarchy of patterns, which are composed of smaller cooperating patterns, which are composed of still smaller cooperating patterns, and so on. When we look at the real world, we do indeed find patterns of patterns of patterns. And they tend to be grouped on the basis of how they cooperate with each other.

It should not be a surprise, then, that, in the process of thinking, we tend to automatically subdivide large patterns into smaller and smaller constituent patterns until we get to indivisible elements. This is, of course, the well-known principle of divide and conquer, which applies equally well to engineers as to battlefield tacticians. I suggest it in fact applies to all intelligent thinking.

A critical component of intelligence is the ability to mentally decompose patterns of things into useful sub-patterns with which we are already familiar. For instance, to understand how an electric coffee percolator works, one needs to take it apart and see what sorts of components cooperate toward its operation. So, let's take it apart. Flipping the coffee pot upside down reveals several screws. Even though the screws may be slightly different from any screw ever seen before, they are quickly recognized as being important sub-components of the coffee pot. They are recognized by their pattern of shape as falling into a class of patterns known as screws. And we know exactly what to do when we encounter screws. We'll later discover that the recognition of how to properly interact with a particular pattern constitutes an *understanding* of it. In this case, we simply need to get and use a screwdriver.

So, another critical component of intelligence is the ability to recognize certain patterns as being members of common classes of patterns, even though the particular patterns to be recognized may not be precisely the same as anything ever seen before. This ability to sort patterns into common categories – referred to as induction – is what allows us to draw certain types of inferences. We'll discuss all these aspects of intelligence in greater detail later. For now, let us stay focused on the concept of cooperation among patterns.

When looking at a car, certain sub-patterns are easily recognized: the wheels, the doors and the dashboard, for example. Each of these may be further sub-divided. For instance, the wheels consist of metal rims, rubber tires and valve stems. I want to claim that many memes cooperate in the formation of a car. But I need to be precise here. A car is not really a meme. For patterns to be memetic they must serve as templates for their own replication or translation, and cars don't typically do that. Thus, I am tempted to think of the patterns of human behaviors, in the making of various sub-components of the car, as the memetic replicators. But, to be completely accurate, the actual memetic patterns that serve as templates for their own replication or translation are likely to be the patterns of neural connections and consequent firings in the brains of people who repetitively execute those certain behaviors.

Fundamentally, neural firing patterns do serve as templates for replication in the process of manufacturing car parts. At the next higher level, the resulting patterns of human behaviors are direct translations of those fundamental memetic patterns. And, at a yet higher level, the car parts themselves, which result from the behaviors, are indirect translations of the base memetic patterns. And so, we may conclude that a car itself is simply an extended phenotypic expression (translation) of many memetic patterns.

Now we are in a perfectly fine position to claim that many memes cooperate in the formation of a car. We may now logically consider the various fundamental memetic actions – the neural firing patterns that direct the behaviors responsible for making all the diverse component parts and then assembling those components – as cooperating toward the overall replicating pattern – the making of a complete car. We'll later see many examples of how memes cooperate with other memes to form bigger and better memes, but, for now, let us return to contemplating living things.

In the context of life, the principle of cooperation applies at all levels of many hierarchical relationships among many sorts of patterns, by enabling them all to better direct energy toward their own perpetuation. Indeed, the hierarchical levels themselves result from the synergistic effects of cooperation. Any two patterns that cooperate toward their mutual perpetuation create a synergistic individuality at a higher level, and thereby gain an evolutionary advantage over other similar but non-cooperating patterns. This critical fact is true by definition: to cooperate means to act toward mutual benefit. And the profound implication is clear: Various sorts of synergy resulting from various forms of cooperation among various styles of replicators represent the very essence of life.

Although it is easy to see how patterns of genes cooperate to form the physical body and brain of a human, it may not be so easy to see that the making of a mature human also requires a cooperative contribution from many more patterns – the behavioral patterns of cultural memes. Indeed, many cultural memes absolutely do cooperate with genes toward their mutual perpetuation. So, let us now consider the sort of cooperation that is typical between genes and memes in the complete formation of a mature human being.

Nature versus Nurture

Is it our genes that cause us to show gratitude for someone's help, or is it our memes? Are we genetically programmed to be grateful, or are we taught by our parents to return favors? These questions, and many similar ones, are at the heart of a long-running debate over where to draw the boundary between behavior that is innate – inspired by genes – and behavior that is learned – inspired by memes.

There are a couple of methods by which we might determine the proper attribution of behaviors to either nature or nurture. One way is to determine which set of replicators – genes or memes – stands to benefit. Just as criminal investigators sometimes identify suspects by following the money trail – that is, by assessing who stands to benefit – so can we follow the 'money trail' for replicator perpetuation. After all, we must be genetically programmed to automatically act in the interests of our genes. And, we must be memetically programmed to act in the interests of our memes. But it is rarely clear which is more responsible for a given behavior.

So, which set of replicators benefits from the expression of gratitude? It turns out that both genes and memes benefit from such behaviors, as reciprocal expressions of gratitude, that tend to cooperatively unify multiple individuals by way of a friendship. As cooperative people survive and have babies, they propagate both their genes and their cultural memes to those babies. So, both sets of replicators benefit. We simply can't know for sure which is more responsible. The best we can say is that both cultural memes and genes may have co-evolved to encourage cooperation through reciprocal expressions of gratitude.

Another way of deciding whether a behavior is either caused by nature or nurture – genes or memes – is to ask whether the behavior would likely be performed by an individual, or group of individuals, raised in isolation, without any exposure to cultural memes. That is, if we could take a bunch of newborn babies, hook them up to machines that would nourish them until they could nourish themselves, and place them on a deserted island, what behaviors would result from their innate genetic programming alone? Unfortunately (from the viewpoint of scientific discovery), no such experiment has ever been done, and none are likely to occur anytime soon. So, we can only speculate on the results. But if we try to speculate on what would happen, we find our views obscured by the very thing we are trying to ascertain.

Unfortunately, we each carry around a lot of cultural bias in our respective thinking. It is simply impossible to separate out that bias. Our ability to think depends on it. How can we possibly know what it would have been like to have grown up without language, for instance? How can we possibly speculate on the boundary between nature and nurture by first supposing we had never been exposed to those terms?

Because of our inherent bias, we must guard against falsely assuming that, even without the ability to mimic others, we humans would just figure out how to do all the things we commonly do. For example, if a human were raised without the benefit of ever seeing another human walk upright, would it figure out on its own how to stand and stroll, or would it remain on all-fours for its entire life? We simply can't know unless we run the experiment. But most people just automatically assume that the act of walking is a natural thing for a human to do. It is such a common activity among all humans that we are biased to think it would happen even without the benefit of mimicry or teaching by parents.

When instances of life make things – call them artifacts – the corresponding activities can be either inspired by genes or inspired by memes. We need to draw a distinction between the two sources of inspiration. For instance, a bird's nest, a spider's web and a beaver's dam are all examples of what Dawkins refers to as *extended phenotypes*. That is, they are natural products, respectively, of the bird's genes, the spider's genes and the beaver's genes – they are indirectly inspired by the genes. While a bird results *directly* from its genes, any nest the bird builds results *indirectly* from the bird's genes. A mature bird knows how to build a nest without ever having seen one built.

Any sorts of behaviors or artifacts that are not inspired by genes must result either by pure chance or by inspiration from memes. We must be careful to recognize that almost all human artifacts are inspired by memes, not genes. I once attended a lecture by a well-known philosopher during which he cited some examples of extended phenotypes: a beaver dam, the Hoover Dam, a spider web and a power grid. He was wrong to include the Hoover Dam and the power grid as extended phenotypes. They are phenotypes of memes, not extended phenotypes of genes. A human, or group of humans, born and raised without exposure to modern artifacts, is unlikely to construct anything like the Hoover Dam or a power grid. There is nothing in the genes of humans to inspire those sorts of construction activities. Big dams and power grids can only be memetic products of societies whose technological memes have already evolved through the development of language, heavy machinery, concrete, electricity, wires, etc.

When it comes to humans, we often can't know exactly where the line is between genetic and memetic influences on behavior. But it really doesn't matter so much that we distinguish between genes and cultural memes, because they are both vertical replicators having very similar agendas of propagation via the proper development of healthy and cooperative individual hosts.

We are now in a position to understand why genes found it beneficial to enable mimicry of behaviors from parents to their children. Independent replicators having similar agendas tend to evolve toward the same goal. It is their common vertical style of propagation that gives human genes and cultural memes their common agendas. They cooperate toward the perpetuation of their human hosts.

Vertically Propagating Memes

Memes that propagate vertically from parents to offspring follow the same propagation paths as genes, and tend to evolve in ways that complement and extend the physical capabilities expressed by genes. Amazingly, vertically propagating memes can even guide the evolution of genes. Surely, memetic behaviors involving the use of such things as clubs and spears must have guided the biological development of strong and reliable opposable thumbs. Any hominid not having an opposable thumb sufficiently developed to grasp a spear would have quickly succumbed to another hominid having such a thumb. Vertically propagating memes of culture can be considered almost as virtual extensions to the vertically propagating sets of genes. They cooperate toward their mutual perpetuation.

Early hominid parents surely taught their children many behavioral tricks for finding food, treating wounds and avoiding danger. The genes benefited from those behaviors. Certainly, we modern humans teach our children all sorts of behaviors for coping with the world – language, social mores and good hygiene. A modern child would be severely disadvantaged without its memes of common culture. Even the act of teaching cultural behaviors to our children is itself best viewed as a behavioral meme that we learned from our parents.

The passing of knowledge in early hominids from parents to children probably did not initially happen so much as a result of parents teaching, but instead, as a result of children mimicking. Genes needed only to build brains that were capable of automatic mimicry, and the evolutionary process, applied to memes of culture and technology, did the rest. Even today, young human brains continue to wire up their neurons for many years after birth by automatically mimicking the behaviors of parents. This simple mechanism enables culture to advance from generation to generation, almost as if it were encoded in the genes.

Vertically propagating memes – passed from parents to children – tend to be either beneficial or inconsequentially inert, because, if they are detrimental, then they adversely impact the survival of their hosts, on which they rely for further propagation. The same is true, of course, for stretches of DNA. Vertically propagated DNA must either code for useful proteins or be inert. As I previously mentioned, most of the DNA in a human's chromosomes are junk filler that has merely hitched a ride using the copying machinery that duplicates the useful DNA. The junk is simply inert DNA that has no detrimental effect on its host. As a memetic analogy to junk DNA, many cultural customs and rituals have no beneficial or detrimental effects on their hosts, yet they get passed on through generations, nonetheless.

While the memes of religion often include many such inert rituals, there must have been some religious behaviors that were beneficial to our ancestors, long ago. For example, it is probably not a coincidence that the religious moral principle known as the *Golden Rule* – 'do unto others as you would have them do unto you' – is so similar to the optimal strategy for the game Prisoner's Dilemma – *Tit-for-Tat* – which can be loosely paraphrased as 'do unto others as they just did unto you'. Tit-for-Tat is a good strategy for living, because, just as the game Prisoner's Dilemma rewards cooperative behaviors, so does life.

Among the many religions that spontaneously emerged in the past million years, the ones that preached behaviors having survival value are the ones that were most likely to have survived. And, Christian attitudes probably did not originate from Christ, but rather, evolved over eons of social interactions. For whatever reason, significant events of the time caused people to adopt Christ, who probably expressed the prevalent religious ethics better than anyone else, as the symbol for attitudes that had long been already adopted by many of the society. Therefore, the principles of Christianity probably preceded Christ, and his popularity merely gave them a new name.

It is difficult to see how religious ceremonies yield any benefit at all in our modern environment. I can easily believe that religion formed the basis for group cohesion, eons ago, at a time when membership in a group became a necessity for survival. And I certainly do believe that the moral principles espoused by religions, like the Golden Rule, had the beneficial effect of encouraging cooperation long ago. But we modern humans are now beginning to divorce the principles of morality from the religions that first espoused them. Our philosophical endeavors now pursue moral truths independent of any religion.

Today, cooperation is encouraged by governmental laws that enforce contractual arrangements, prevent crime and facilitate commerce. Since governments now provide the protections and cooperative efficiencies that religions used to provide, religions have been reduced to pure ceremony. Those ceremonies no longer serve the adaptive purpose they used to. Yet, religious people today continue the traditions of their religions simply because their parents taught them to.

Whereas religious memes probably evolved so as to create a cooperative morality, language memes evolved so as to effectively coordinate those individuals whose moral imperatives now drive them to cooperate. Morality provides the motive for cooperation, and language provides the effective means for cooperation.

Vertically propagating memes of cultural are best considered as patterns of behavior that just happen to get themselves perpetuated in a world full of mimicking humans. They are replicators in their own rights, with their own replication agendas. Just as genes selfishly act with no regard for their hosts, except to the extent that the hosts can be used for propagating the genes, we should expect that memes also act with no regard for their hosts, except to the extent that they can be used for propagating the memes. Sometimes human hosts benefit from their memes, but not always. Whereas vertically propagating memes tend to benefit their hosts, we'll now discover a class of memes that are more likely to reveal their selfish natures.

Horizontally Propagating Memes

When children mimic their parents, the result is the *vertical* propagation of culture. When children mimic other children, the result is the *horizontal* propagation of various sorts of fads. Of course, fads are not limited to children, they are a big part of all our lives. For instance, most of us pay attention to what our peers are wearing in hopes of successfully mimicking the latest fashions. Those who don't pay attention to fashion, wearing instead the silly looking wide ties and lapels that were popular in the 1970's, are easily identified and are often ridiculed for such behavior. Why do we make fun of them? We have been programmed by our defining replicators to expect a cooperatively unifying social allegiance from everyone in our society. We effectively encourage group loyalty among our compatriots by socially demoting those who exhibit eccentric or non-unified behaviors.

Fads propagate horizontally. And that style of propagation allows them to include behaviors that are detrimental to the welfare of individuals who propagate them. For example, most high-school principals know that one student's commission of suicide can often inspire others to commit the act, thereby propagating a detrimental behavior among several individuals at tremendous expense to their genes. A great many teenage behaviors, while much less drastic than suicide, are indeed detrimental to normal genetic perpetuation.

In the minds of teenagers, ideas coming from their parents tend to benefit their genetic welfare, while ideas coming from their friends are often dangerous and destructive, benefiting only the perpetuation of the memes that inspire those ideas. Kids encourage each other to drink alcohol, take drugs, race their cars and do absolutely anything in lieu of studying. Parents, on the other hand, encourage their kids to do whatever best prepares them for raising families. This situation is not a coincidence; it merely results from a natural competition between vertical memes and horizontal memes for control of our human minds.

Why do our brains prefer some memes over others? By what criteria do we select the memes we choose to embrace and recite to others? The detailed answer will have to wait, but a short answer goes as follows: The selection of memes by brains is automatically performed on the basis of whatever memes are expected to yield the most happiness, both in the present and especially in the future. The selection of memes by their expected future effects, while enabling the valuable principle of investment, opens the door for memetic control of humans. Consequently, some memes have found ways of getting themselves propagated by giving us false impressions of the future. They make some people believe in fantasies, such as going to heaven with 72 virgins as reward for a suicide bombing. Those improper memetic beliefs sometimes cause us to do things that are not at all in the best interests of our genes. And so, horizontally propagating memes can be extremely detrimental to the propagation of genes.

It is not necessarily the case that horizontal memes must be detrimental to genes. It just happens that, among the many memes that propagate from peer to peer, there are some that aid the propagation of genes and some that are detrimental. When parents come to recognize the few horizontally propagating memes that tend to aid gene propagation, those memes are labeled as 'good' and are adopted for inclusion into culture. Those corresponding behaviors are encouraged by parents and are thus propagated vertically. The horizontally propagating memes that are left over tend to be either detrimental or inconsequentially inert.

Most memetic behaviors that propagate among us are inert junk. Hula-hoops and pet rocks neither help nor hinder their hosts to any appreciable degree. And, most of the postings to the Internet have no value to anyone, yet they are sometimes repeatedly propagated from site to site in the manner of useless urban legends. Unfortunately, as Dawkins points out, it is much easier to create a useless idea than a useful one, so useless ideas tend to proliferate. Tunes, jokes and fashions certainly fall into the category of inert junk memes. Likewise, many of the products we produce don't make us more survivable, nor do they make us more likely to propagate our genes. Trinkets and 'chatchkies', for instance, make some of us happy and are thereby able to get themselves copied. It seems those sorts of junky products have hitched a ride using the very same system of commerce that came into existence for reproducing useful products.

Once a system of reproduction gets going, it just goes and goes. We see it in the perpetuation of DNA and also in the perpetuation of memetic products of commerce. We may analogize the accumulation of junk patterns emerging from a system of reproduction in the following manner. Suppose that, on top of a company's copy machine, is a clever cartoon.

Whenever employees see it they get some happiness from it, which inspires them to make a copy of it. Then, because of minor imperfections in the copying process, the found original is always taken and the new copy is left behind. After many generations of copying, the cartoon begins to accumulate black dots on it – junk patterns. But, so long as the cartoon is recognizable enough to get people to copy it, it is thereby able to direct energy toward its own perpetuation. And, similarly, so long as the junk filler in DNA does not interfere with the meaningful genetic patterns, the important gene patterns will continue to direct the perpetuation of the entire set of DNA, taking the junk DNA along for the ride. And, so long as trinkets and chatchkies don't consume too much of our critical resources they will continue to get perpetuated.

It is getting difficult to classify modern memes as propagating either vertically or horizontally. While we do learn many things from our parents, we learn a great deal many more things in school, from other people's parents. This 'diagonal' style of propagation muddies the waters of our understanding with respect to the effects of memes. Do they work *with* our genes or *against* them? Are they in cooperation, competition or merely co-existence? Indeed, there are many memes of all types, merely exploring various avenues of perpetuation for their own benefit.

All Patterns are Products of Nature

The simple process of evolution, applied to all sorts of replicators, is able to yield enormous complexity. One might easily mistake the complexity of life as an indication of an omnipotent designer, because we typically think of complexity as arising from intelligence. But that is a backward view of the situation. The truth is, intelligence arises from complexity. Interesting things always seem to happen at the "edge of chaos," says M. Mitchell Waldrop in his book *Complexity* (1992).

We'll later find that intelligence likewise requires a frothy process of trial and error to produce its valuable fruits. The complexity produced from the simple process of evolution provides the many degrees of freedom required for explaining things that appear to some as incredible, miraculous, or divinely inspired.

Evidence shows that we humans have evolved to be deterministic biological machines, automatically seeking cooperative relationships and comfortable situations that are likely to be conducive to the perpetuation of our defining replicators. This description tears down any boundary between ourselves and the rest of nature. We are indeed as natural as anything else.

We humans are what we are because nature prefers for life to be that way. Nature endorses humanity as the dominant species because we humans are closer than any other species to the style of life preferred by nature. We are compassionate because nature has arranged so that expressions of compassion among the members of a group tend to enhance that group's ability to survive. And, as the group survives, so do its members. We are intelligent because nature gives statistical preference to life that can discover better ways to perpetuate its genes. Morality combines with intelligence in humans to produce a species that is likely to engage in many forms of synergistic cooperation. We don't volitionally choose to be moral any more than we choose to be intelligent. We just are – at least, many of us are.

Chapter 3. **Genes and Memes**

I have already alluded to several obvious similarities between genes and memes. It turns out that there are many more subtle similarities between genes and memes that can only be appreciated by rigorous comparison. So, this chapter will focus on identifying those similarities. By looking closely, we find that many characteristics of genetic evolution have loosely analogous counterparts in the realm of memes. By using what we know of genes to elucidate the analogous memes, we can gain some insight into exactly what memes are, why they evolved, and how they enable us to think.

This is a somewhat difficult chapter to read. It necessarily jumps all over the place, citing disparate examples of gene-meme commonality. And, to someone already familiar with memes, much of the information may seem trite. Yet, there are some nuggets of information here that are commonly misunderstood, but are critical for understanding intelligence. In particular, it is absolutely crucial that we correctly understand how memes mutate. Many authors incorrectly cite typographical errors, and similar sorts of random events, as the primary mechanisms by which memes mutate. That is completely inaccurate. The primary mutation mechanism for memes becomes crystal clear when we fully consider how genes mutate. Further, it is extremely important to understand the critical role cooperation plays in both genetic and memetic evolution.

Finally, the Lamarckian nature of memetic evolution appears to be unlike anything in the domain of genes, and is critical for understanding how human intelligence efficiently transcends the intelligence of random biological evolution. But we will discover that Lamarckian evolution is determined more by the environment than by the types of replicators, and that genes will indeed eventually evolve by Lamarckian means.

Fundamentals of Genetic and Memetic Evolution

There is a translation mechanism by which the information in human genes ultimately gets converted into flesh and bone. The mechanism is said to *express* the genetic information by creating a phenotype – a physical body. The same is true for memes. Memetic information can also be translated or expressed into various sorts of physical products or environmental effects. So, a meme too can produce a phenotype – a physical product. Just as every biological species is described by its own string of genes, so does every manufactured widget and every repetitively-performed service that we humans routinely produce have a corresponding execution recipe, or string of memes. Let us now consider exactly how genes express themselves.

Genes are simply stretches of DNA – long strings of special molecules called nucleotides. There are four types of nucleotides that can be arranged in any order to compose a stretch of DNA. We refer to those four different nucleotides by the abbreviations: A, C, G, and T (for adenine, cytosine, guanine and thymine). So, a given gene might have a form such as: TTGCACTGATGACCG...

A string of nucleotides is actually arranged in the form of a ladder, such that each rung on the ladder is a pair of complementary nucleotides. If there is an A on one side of a rung, then there will be a T on the other side of the same rung. A's and T's are complementary. C's and G's are also complementary. If one side of the ladder has the string TTGCA, then the other side will have the complements AACGT. When a string of DNA replicates, it happens as a result of the close proximity of certain enzymes that automatically 'unzip' the two complementary halves. We might visually imagine the unzipped strings as having 'dangling' attraction sites to which new complementary pieces can be attached. The truth of the situation isn't quite that simple, but this simplified visual model will serve our purposes well. So, let us further imagine that, as the appropriate complementary pieces float by, they are automatically attracted and bound to the respective dangling attraction sites, thus re-forming the complement that just split off.

Genes physically express themselves according to their encoded recipes by chaining together naturally occurring molecules, known as amino acids, into much bigger and more complex molecules known as proteins. Those proteins then automatically connect together to form all the different types of biological matter.

The sequence of nucleotides in the pattern of a given gene, taken in sets of three, describes the sequence of amino acids to be strung together to form a unique style of protein. A set of three sequential nucleotides forms a *codon*, which specifies a particular amino acid. The cellular machinery simply reads the letter sequence in the DNA, gathering and assembling the specified ingredients into a corresponding string of amino acids. That string then folds itself up in a very precise manner to form a protein – a bumpy sort of glob. The differences between various proteins come primarily from their different sizes, shapes and locations of charged attraction sites on their lumpy bodies. Proteins can have protruding bumps and sunken valleys on their surfaces. The bumps and positive attraction sites of one protein may or may not fit snugly with the valleys and negative attraction sites of another protein. Proteins that fit well together form a tight bond that can be very stable and can lead to the construction of biological matter. We may say that such proteins have complementary interfaces.

The constant jostling of molecules in cellular liquid causes proteins to drift about in somewhat of a random manner until they find complementary mates with which they can connect. It is easy to see how critical it is that life be poised within a temperature range on the edge of chaos. If the energy falls below a certain level, cells freeze up and the constant jostling of molecules ceases. If life is too energetic, the chaotic motion of particles causes long and delicate molecules to break apart soon after they are formed. In an analogous manner, memes likewise thrive at the edge of chaos. Memetic creativity depends on islands of stability – known truths – surrounded by frothy waters of wildly diverse opinions, so long as they are not too wild.

Luckily, the genetic process of folding up a string of amino acids into a protein is extremely reliable. However, there are some proteins in which folds can sometimes flip the wrong way, even long after the construction of the protein is complete. To illustrate the problem with this scenario, imagine a protein that folds itself into a particular style of glob with a hinge point that always tends to fold in one direction just before the protein is complete. If the hinge point were to fold in the opposite direction, then the shape of the glob would be very different. Such precarious proteins exist, called *prions*, and they form the basis for *mad cow disease*. The disease is contagious due to the very unfortunate fact that the improperly folded proteins are able to catalyze the same improper folding in other properly folded proteins of the same genetic coding.

Prions spread their improper style of folding whenever a properly folded protein flips at its hinge point simply after 'bumping into' an improperly folded protein of the same sort. Prions can remain stable for very long periods of time in the carcasses of dead animals, and can spread their improper folding to normally folded proteins in healthy animals after having been ingested. So, cows, and even humans, can catch *mad cow disease* simply by eating the meat of a cow that had the disease.

Is there a memetic counterpart to prions? Yes, in a very loose sort of analogy there is. Prions are a bit like 'double-entendres'. Just as prions can flip from one configuration to another, producing two different shapes from the same genetic sequence, so can a pun cause a flip between two vastly different meanings for the very same single sentence. For example: "Is life worth living? It depends on the liver." Luckily, puns aren't very contagious. Don't worry; we'll discover much more meaningful similarities soon.

Back to genes. The tens of thousands of genes in a human's DNA are grouped into 23 sets, called chromosomes. Each gene in a chromosome codes for a particular style of protein needed by the body. So, there are many slots in each chromosome, each holding a specific gene that codes for a particular protein. When we look at the same slot over many different humans, we find there are often several slightly different versions of genes performing the same genetic function. Two different versions of the same gene, competing for the same functional slot in a chromosome, are called *alleles*.

A similar concept applies to memes. For example, in a typical kitchen we are likely to find a stove, which can be of several different styles. We might find a gas stove, or an electric stove with exposed burners, or an electric stove with a glass top. All these different styles of stove are intended to perform the same function, and they are therefore analogous to alleles, but they exist as alleles in the domain of memes. Note that the allelic memetic replicators are not really the stoves themselves, but the more fundamental patterns that get themselves replicated in the making of stoves (whichever patterns tend to serve as templates for their own replication or translation). The stoves themselves are the phenotypic expressions of the memetic replicators.

Back to genes. Most of the trillions of cells in a human body come into existence by *mitosis* – the normal process of cell division in which the descendent cells each receive all the chromosomes of the parent cell. Even though it only takes 23 chromosomes to describe all the genetic proteins of a human, a typical cell contains 46 chromosomes, 23 from the individual's mother and 23 from the father. The mother and father each contribute a full set of functional genes. So, a typical cell holds two codings for every protein needed by the body. For any given gene slot, the two available codings can be slightly different, making them alleles.

In addition to mitosis, there is another special type of cell division, called *meiosis*, that takes place in male testes and female ovaries. In the meiotic process of cell division, a cell containing all 46 chromosomes splits off sex cells – *gametes* – each containing 23 chromosomes. These sex cells receive half their genetic material from the mother's chromosomes and half from the father's chromosomes. The process of choosing whether to take a particular stretch of DNA material from the mother's chromosomes or from the father's chromosomes is called *crossing over*. This random process of reducing the two functional sets held in the 46 chromosomes down to a single set of 23 causes each of a male's sperm cells and each of a female's egg cells to be a unique combination of the individual's maternal and paternal DNA. I'll clarify this with an analogy in the realm of memes.

Imagine that each person has two books, book A and book B, embedded in their genes. Both books discuss the same topic, and are written by different authors but from the same outline. Each paragraph of book A deals with the same concept as the corresponding paragraph of book B. But the different authors use different words within the corresponding paragraphs. And while two corresponding paragraphs deal with the same topic, their contents sometimes take slightly different meanings. The meiotic process creates a new book by taking some pages from book A and some from book B, switching between the books – crossing over – at intervals that are mostly random. The resulting book is quite readable and understandable even though it is different from both book A and book B.

Similarly, it is common for book authors to pull quotes from other books as support for their line of reasoning. Such acts are real-world examples of a meiotic style of memetic recombination, analogous to genetic recombination through 'crossing over'.

Similarities abound between hierarchical groupings of DNA in the domain of genes and hierarchical groupings of text in the domain of memes. Nucleotides loosely correspond to letters, codons correspond to words, genes correspond to paragraphs, and chromosomes correspond to chapters. Just as codons are symbolic of amino acids in the genetic representation of proteins, so too are some words symbolic of objects in the real world. And, just as all 23 chromosomes describe how to construct a human, so can the chapters of a book describe how to construct an artifact such as, say, a coffee percolator.

There seem to be many similarities between genes and memes. Kate Distin (2005) even recognizes a similarity between recessive genes that are held but not expressed, and 'recessive' memes that are known but not believed. For example, if the set of genes held by a particular individual contains both a gene for blue eyes and a gene for brown eyes, the recessive blue-eyed gene will have no effect on the phenotypic expression of that human. Similarly, the memetic concept of ghosts can be known to an individual, but unless it is believed it will have no effect on the behavior of that individual.

We should not be surprised to find so many similarities between genes and memes if indeed it is true that, fundamentally, they are the very same things – patterns that serve as templates for their own replication or translation. The many similarities between genes and memes should make us suspicious that memes, like genes, have their own agendas for getting themselves replicated. This idea takes on real significance when we admit that our minds are likely to be mostly deterministic. Such a meme-centric viewpoint allows us to see why we humans tend to replicate some behaviors and ideas that tend to serve no purpose other than to further the replication of those very same ideas.

Replication is only one part of the evolutionary process. We have a much more important lesson to learn from the ways in which genes and memes similarly mutate.

Mutations of Genes and Memes

Patterns are only able to evolve if they sometimes experience subtle changes that can be propagated to descendants. In the genetic realm, we call those changes 'mutations'. There are two distinctly different styles of mutations that can occur to the patterns of DNA held in the chromosomes. And, as it turns out, they both similarly apply to memes. The most important of these is often overlooked, especially in the realm of memes. I'll describe both, starting with the more commonly recognized style of mutation.

With respect to DNA, the style of mutation with which we are most familiar occurs when a very rare mistake in the copying process, perhaps caused by an errant cosmic ray, replaces one nucleotide with another. This style of mutation is referred to as a *single nucleotide polymorphism*. In the context of our book analogy, this style of mutation corresponds, in the domain of memes, to something like a typographical error that changes a single character of text.

I want to make a subtle but important point here. By changing a single letter in a genetic sequence, we will see a change in the amino acid at the corresponding point in the string of amino acids composing the protein. But, as the long string of amino acids folds itself up into its bumpy shape, the new shape will not necessarily be largely different from the shape it would have been without the change. Therefore, a single mutation applied to a stretch of DNA may result in a very minor change to the shape and corresponding functionality of the resulting protein. Through this mechanism of graceful and subtle change, biological adaptation may proceed in very small steps toward optimization.

The ability for subtle change is a critical characteristic of successful evolution that we are having trouble duplicating in the memetic domain of evolving software. Any good programmer can tell you that even a single character change in the source code of a program usually renders the entire program absolutely useless. So, the secret to producing software capable of evolving hinges on discovering how to cause software to mutate gracefully. I believe we will soon discover how to build a layer of abstraction on top of our machine codes that will enable software objects to mutate gracefully and thereby evolve incrementally. Now, let us move on to the second and more important style of evolution.

There is indeed a different style of mutation that might enable a graceful evolution of all sorts of memetic patterns, including software. It is a style of mutation that, as happens with genes, involves the shuffling of whole functional units through something like the genetic processes of meiosis and 'crossing over'. This other style of mutation is extremely important in the domain of genes, yet is often overlooked in the domain of memes.

Humans don't evolve so much by mutation as by permutation. Almost every newborn human baby has a brand new pattern of genes that has never previously existed, but all those unique gene patterns come primarily from combining different permutations of well-established alleles, not from random nucleotide polymorphisms at the DNA level. It is generally true for most functional sorts of composite patterns that, by combining some functional components from this pattern and some functional components from that pattern, you get a whole new pattern that is much more likely to be functional than if it were randomly mutated at a lower level, within the components. Such is the recombinant process by which both genes and memes mostly evolve.

In the memetic domain, there are many obvious examples of this style of mutation. For example, businesses routinely combine common memetic components in different ways to produce new products. Transistors, capacitors, wires, screws, hinges, knobs and buttons are but a few of the many components that get used over and over again, combined in various ways, in the products we humans routinely manufacture.

Another memetic counterpart to the genetic shuffling of genes is evident in the way scientific ideas evolve through books. I previously alluded to the idea that most scientific books present quotes from many other books as means of support to the central argument. Perhaps I could have composed this entire book by simply combining snippets from many other books. It may not have been quite as readable, but it could have easily contained all the pertinent information. A typical book consists of many pre-existing ideas, or memes, combined in some new way, just as a typical human is built from many pre-existing genes combined in some new way. Sentences combine pre-existing words in new ways. Songs combine pre-existing notes and chords in new ways. It seems that memes mutate mostly by *recombination.*

Recall the problem of creating software that successfully evolves. Instead of mutating the software by single character substitutions, suppose we were to collect all the coded subroutines and functions that exist today and combine them in all possible ways. There are surely millions of such subroutines – chunks of computer code that perform specific functions. And, with no upper limit on the number of subroutines in a combination, there are an infinite number of combinations. But it is easy to see that among those many combinations are all the software programs that we routinely use. And, I find it interesting to speculate on the range of useful products that could emerge from such a powerful means of memetic evolution. Could we possibly discover a program that could converse in English through this technique? An even more critical question is: will we want to construct evolving software? By allowing software to evolve, we lose control and understanding of it. It takes on something of a life of its own.

In both the genetic and memetic realms, as the number of alleles grows, the number of possible allelic combinations explodes. A quick and dirty estimate of the possible number of allelic combinations, for genes, exceeds an amount expressed in decimal notation as a one followed by 300 zeros (assuming two alleles for each of a thousand genes). As a frame of reference, our world's current population is less than an amount expressed as a one followed by 10 zeros. In fact, the number of particles in the universe is estimated to be an amount expressed as a one followed by about 80 zeros.

As wildly expressive as DNA is, human language is far more so. Whereas DNA uses only 4 nucleotides, English uses 26 letters. Whereas codons can express only 64 different amino acids (only about 20 exist), there are tens of thousands of English words. The expressive power of the English language is much greater than that of genes. Human language allows us to conceive of abstract things that would just be impossible to imagine without it. For example, how could we ever conceive of radio waves or black holes without a conversational and mathematical language in which to express them?

In the spirit of mixing and matching various working components in new ways, recognize that biological evolution is very good at re-using component patterns and re-applying them to novel uses. Goose bumps are an interesting example. We may logically speculate that the original value of goose bumps came from their ability to increase insulation. I get goose bumps when I am cold simply because a hairy ancestor of mine also got goose bumps when he was cold. His goose bumps were adaptive because they caused his thick hair to stick out perpendicular to his skin. Instead of laying flat against the skin, the puffed out hair was able to hold more air molecules in such positions that they would provide a layer of insulation.

I also get goose bumps when I am frightened, and I may assume it is because some hairy ancestor of mine also got goose bumps when he was frightened. It must have been a strange quirk of wiring in one ancestor's brain that caused the mechanism designed for increasing insulation to be applied when frightened. The quirk of mis-wiring, lo and behold, had adaptive value. By causing his hair to stand on end, the goose bumps had the side effect of making him look bigger. His more-imposing size must have been enough to cause some opponents to flee rather than fight. It is, to be sure, a subtle benefit. But over the course of many applications, even a subtle benefit will statistically emerge from the morass of all possible gene combinations. Evolution works in mysterious ways.

I made this connection in my own mind, between the two different triggers for goose bumps, after seeing a bird puff up its feathers during a confrontation with another bird. Isn't it interesting how ideas combine in the ways they do? My memetic ability to *see* similarities in patterns of action is the complementary flip-side to the genetic ability to *use* similar patterns of actions for different purposes. What I am trying to say is that intelligence must involve patterns of patterns of patterns, both in its ability to understand complex objects and in its ability to create complex objects. It should not be a surprise that an act of understanding breaks patterns into recognizable sub-patterns, if indeed it is the case that an act of creativity amounts to combining functional sub-patterns into bigger and more complex functional patterns.

The recombination of genes through meiotic shuffling is a wonderful source of subtle genetic mutation. And, as I previously mentioned, its counterpart in the memetic realm is often overlooked. Consider this passage from Steven Pinker in which he obviously dwells on mutations in the style of typographical errors as the only mechanism for mutating memes. He intends to sarcastically support his belief that memes simply cannot account for the wonders of human intelligence.

> A meme impels its bearer to broadcast it, and it mutates in some recipient: a sound, a word, or a phrase is randomly altered. Perhaps, as in *Monty Python's Life of Brian*, the audience of the Sermon on the Mount mishears "Blessed are the peacemakers" as "Blessed are the cheesemakers." The new version is more memorable and comes to predominate in the majority of minds. It too is mangled by typos and speakos and hearos, and the most spreadable ones accumulate, gradually transforming the sequence of sounds. Eventually they spell out "That's one small step for a man, one giant leap for mankind." ... Natural selection designed the mind to be an information processor, and now it perceives, imagines, simulates, and plans. When ideas are passed around, they aren't merely copied with occasional typographical errors; they are evaluated, discussed, improved on, or rejected. Indeed, a mind that passively accepted ambient memes would be a sitting duck for exploitation by others and would have been quickly selected against.
>
> **– Steven Pinker**, *How the Mind Works* (1997, pp.209-210)

Pinker completely overlooks the most important style of mutation applying to both genes and memes. When ideas are passed around, they are combined in various ways to create bigger and better ideas. Such is the 'meiotic' style of memetic mutation that enables the evolution of technology and thereby provides tremendous adaptive advantage. Pinker is absolutely right that "natural selection designed the mind to be an information processor." But information is always composed of recurring patterns, and recurring patterns are memes. Within several chapters, we'll see exactly how the mind "perceives, simulates, and plans" through the automatic manipulation of memes. And, we'll discover the memetic algorithms by which ideas are "evaluated, discussed, improved on, or rejected."

Among the various authors who discuss memes, Kate Distin, in her book *The Selfish Meme* (2005), is the only one I recall recognizing the critical distinction between the two styles of mutation I cite here: (1) the alteration of replicators, and (2) the recombination of replicators. She correctly recognizes that, in order to be successfully recombined, memes must be *particulate* – they must have a clear structure with well-defined boundaries that remain stable through repeated recombination. Indeed, I will later show exactly what the particulate neural structure of a meme is and what its clear boundaries are.

I'll soon show that the combining and recombining of memes is typically done in a more directed and methodical manner than the random shuffling of genes. The method by which memes are combined makes human cognition and technological innovation happen at a much faster rate than biological evolution. That directed method is possible, in part, due to the Lamarckian nature of evolving memes, which we will address by the end of this chapter. For now, let us continue to examine the similarities between genes and memes.

Competition Among Replicators

Now that we have considered how various patterns tend to replicate and mutate, we are left with studying the final ingredient in the process of evolution – the competitive process of selection. When we think of evolutionary competition among sets of genes, we typically conjure up the slogan "survival of the fittest" and an image of nature as "red in tooth and claw." Brutal assaults are endorsed by Mother Nature as she coldly doles out the reward of survival to the strong, by encouraging them to savagely eat the weak.

Such a perspective, while certainly true for lower forms of life, completely ignores the fact that life tends to become more civilized, organized, intelligent, compassionate and moral, as it evolves. Perhaps it is true for all instances of planetary life, just as has happened here on

Earth, that the brutal forces of competition and selection will always yield compassionate morality and intelligence as their products. Indeed, I will elucidate such an arrow of evolution quite clearly in a later chapter. But, for now, let us continue our methodical understanding of evolution. To better understand natural selection, we will examine some of the different styles of competition among replicators, again, noting similarities between genes and memes.

We start with competition between genes of the same species. I have already mentioned alleles competing for the same locus in a chromosome. And I have already indicated that the same sort of competition is evident between memetic alleles. Recall that various styles of stove compete for a spot in a new kitchen. There is another style of competition between genes of the same species, having to do with competing *sets* of genes, each of which are good for a different strategy of survival. This concept will be much easier to convey when we talk about how sets of genes and sets of memes cooperate toward a particular perpetuation strategy. So, I defer the discussion until the next section.

The sort of genetic competition with which we are most familiar is a style that crosses boundaries of species. For example, a gene that is involved in an antelope's ability to run fast can compete with a gene that is involved in a cheetah's ability to run fast. Whichever gene does its job better is the gene that will prosper at the other's expense. Let us briefly consider how the competition between genes of cheetahs and genes of antelopes might have guided their respective evolutionary developments.

At some point, cheetahs evolved powerful jaws and sharp teeth with which to eat large animals, such as antelopes. This development in cheetahs likely caused antelopes to develop a coloring that camouflaged their bodies against the typical landscape background. But then, cheetahs that were able to find food and survive were the ones whose visual acuity and depth perception became more sensitive to the motion of antelopes, say, through stereoptic vision that required both eyes to face forward. In turn, perhaps antelopes grew yet longer legs for running even faster. This back and forth nature of competition between various genes of different species is sometimes referred to as an *arms race*.

There are also arms races of memes. In fact the name *arms race* comes from an actual competition between memes having to do with military weaponry. So, clubs competed with spears, spears competed with guns, guns competed with cannons, and so on, all the way up to nuclear missiles now competing with missile defense systems. Each stage of weaponry can be considered as loosely analogous to a memetic *species* of weapon.

Surprisingly, the fundamental notion of competition between replicators is independent of the various substrates in which they exist. Indeed, a replicating gene can just as easily find itself in competition with a meme as with another gene. In fact, any pattern can find itself in competition with any other pattern, independent of substrate, if the perpetuation of one pattern tends to diminish, in whatever manner, the likelihood of the other pattern's perpetuation. For instance, the memetic use of birth control pills is a clear example of direct competition between a behavioral meme and various patterns of genes. Indeed, the memetic use of birth control pills diminishes the likelihood of genes getting themselves perpetuated.

Given that replicator competition can cross substrate boundaries, we are forced to conclude that the environment of a specific gene includes, not just the other genes within the same individual, but also all the genes and memes of all other proximal life forms as well. Indeed, the environment of any replicator is primarily determined by whatever other replicators are able to exert a phenotypic influence in the local area. Of course, the definition of 'local' is changing dramatically as communication networks now allow phenotypic influence to instantly spread clear around the world.

And so it is with life: A constant battle between one set of genes and another set of genes, between vertical memes and horizontal memes, and even between memes and genes. All replicators selfishly fend for themselves, but they also constantly search for allies with whom they can cooperate. And once in a while, alliances emerge that have great staying power.

Cooperation Among Replicators

For a given instance of a gene to be deemed fit, it first must be compatible with the other genes in the same set of chromosomes in which it finds itself. Therefore, as a gene descends through the generations, it is constantly 'looking' for allies in the form of other compatible genes, while at the same time competing with similar genes, alleles, for its particular slot in the chromosome. Let us explore this extremely important idea that genes can *cooperate* toward their mutual perpetuation.

Some sets of genes tend to work well together, and they tend to travel through the descendent population together. They are said to be *co-adapted*. As an example, I have often wondered about the development of long necks on giraffes and how they must have required compatible development, or co-adaptation, of large lung size. If the length of an animal's neck is increased willy-nilly, it will eventually reach a point where the volume of air in the trachea is greater than the volume of air in the lungs. At that point, with every breath, the animal would inhale the same air that it previously exhaled. It would get no oxygen in the process of breathing. We may logically speculate that genes for long necks and genes for large lungs in giraffes would likely be big losers separately, but are clear winners when acting together. They certainly do cooperate toward their mutual perpetuation.

As another example, consider that cheetahs have evolved a set of co-adapted genes that are consistent with their strategy for eating meat. Meat eaters need a digestive system that can process meat; they also need the speed to catch it and big teeth to kill it. Antelopes, on the other hand, have evolved a set of co-adapted genes consistent with their strategy for eating grass. Grass eaters need to be able to digest grasses, and they likewise need the ability to escape from the meat eaters. Grass eaters usually have eyes on the sides of their heads, giving them a wider field of view, while meat eaters have eyes on the fronts of their heads, enabling stereoptic vision for better hunting and chasing. There is an old expression: "Eyes on the side, like to hide; Eyes in front, like to hunt." The important point here is that it takes a *set* of cooperating genes to implement a strategy for survival.

There are similar effects of co-adaptation in the realm of memes. Consider the way computer hardware and software adapt toward compatibly exploiting a particular market niche. In the 1990's, Intel's microprocessor, along with the Windows operating system, tended to exploit common desktop systems, whereas Apple's Macintosh computer, along with its proprietary operating system, specialized in graphics applications, and Sun Microsystem's Sparc microprocessor, along with the Unix operating system, were the traditional choice for servers and engineering workstations. In each case, the hardware evolved so as to become faster for its own particular niche, and the software evolved to take advantage of the specialized increases in speed. Through generations of hardware and software, each co-adapted pair has become more specialized toward its own niche (although, this is a rather poor example because the three niches aren't highly differentiated in terms of their required processing capabilities).

Back to genes. While it is fairly obvious how genes within a species can co-adapt, it may not be so obvious that genes from two vastly different forms of life can also co-adapt. Actually, co-adapted genes of different species are typically referred to as having *co-evolved*. As an example of co-evolved genes, consider again the genes in flowering plants that determine their visual characteristics and the genes in bees that help them find nectar. Plants have evolved visually distinctive flowers and sources of nectar so as to attract bees for the purpose of spreading the plants' pollens. Bees have evolved eyes attuned to finding flowers that are likely to supply nectar. As I previously mentioned, the genes of bees and the genes of certain plants cooperate toward their mutual perpetuation.

As another example, consider how we normally think of a human body as being completely defined by its chromosomes. That's not quite true. In addition to the chromosomal DNA in the nucleus of each cell, humans also have another small amount of critical DNA inside the mitochondrion, which exists also inside each cell. It is now believed by many scientists, thanks to the brilliant work of Lynn Margulis (former wife to Carl Sagan), that the mitochondrial DNA came to exist alongside the chromosomal DNA as a result of a merging between two separate organisms. Those organisms apparently coexisted in some sort of a symbiotic relationship long ago. Indeed, our physical bodies are phenotypic expressions of two sets of DNA, chromosomal and mitochondrial – two sets of patterns working together to achieve greater survivability for both than would be possible separately. As I previously stated, such synergistic cooperation is a critical theme that I believe to be the essence of all life.

Consider the mutually beneficial relationship between our human bodies and the E. coli bacteria that help us digest food. Our human genes depend on the genes of the bacteria for aid in digestion, and the genes of the bacteria enjoy a secure existence in our intestines – a match made in heaven. Both sets of genes, human and E. coli, cooperate toward their mutual perpetuation.

Amazingly, the grouping of genes within an individual (plant, microbe, animal or human) is of secondary importance to the grouping of genes by the metric of cooperation, wherever they may exist. It just turns out that genes within any given individual tend to be highly co-adapted, and thereby, extremely cooperative. The evolutionary phenomenon of co-adaptation is best viewed as an evolution toward better cooperation.

Cooperation is not restricted to replicators within the same substrate. Consider that genes and memes certainly cooperate toward their mutual perpetuation. The memes of vertically propagated culture have evolved so as to aid the perpetuation of our human genes, because those memetic replicators rely on humans for their own perpetuation. And the memes of technology have evolved so as to aid the perpetuation of our human genes, because we humans tend to select whatever memes of technology happen to make our lives happier, healthier, easier and safer. Benefit flows in the opposite direction as well. Our human genes have evolved the capability of mimicry to aid the perpetuation of cultural and technological memes. Both sets of replicators, genes and memes, tend to benefit from the various relationships between them, whenever they cooperate toward their mutual perpetuation.

Just as a computer relies on the cooperating elements of hardware and software, so does a human rely on similar sorts of cooperating elements contributed by both genes and memes. Our genes build us and our memes move us. They cooperate toward their mutual perpetuation, even though they exist in completely different substrates. Notice that I have lumped vertical memes of culture into the bag of replicators from which mature humans are constructed. We might as well throw mitochondrial DNA and the genes of E. coli into that bag as well.

There are all sorts of diverse replicators that come together to form a mature human. The better they all cooperate toward their mutual survival and replication, the fitter the human. There are all sorts of memes that aid human genes in their blind quest for perpetuation. Mimicked behaviors of language, of hygiene, of gathering food, and of manufacturing weapons, all aid in genetic survival. And genes, by implementing mimicry, certainly facilitate the perpetuation of all memes, even though some memes have found ways to get themselves perpetuated without providing any benefit to the genes. Notwithstanding those devious, parasitic, horizontal memes, there is indeed a great deal of cooperation occurring between various memes and various genes.

The effects of cooperation among replicators can be most easily seen in the realm of technology memes. Many different technological procedures cooperate in many various ways to construct many varied sorts of products. In fact, much of what we think of as intelligent creativity arises from the combining of two or more memes to create a bigger meme. The constituent memes can be said to cooperate if they create a better meme.

When memes combine, they start an evolutionary process of improvement through co-adaptation. For example, the invention of the automobile simply combined the idea of a carriage with the idea of an engine. And ever since then, engines have become ever more intimately associated with the carriage. The association has forced both components to evolve in ways that reinforce the cooperation between them. In all cases, replicator co-adaptation tends to result in increased synergy of cooperation.

All sorts of memetic patterns now cooperate with each other toward their mutual perpetuation. Think of the procedure for making a *wheel* and the procedure for making an *axle*. Think of the memetic recipe for creating a *bolt*, and the memetic recipe for creating a *nut*. Together nuts and bolts achieve tremendous synergy, as do wheels and axles. Yet, apart they are worthless. Think of high-speed communications lines and how they cooperate with electronic routers to form the Internet. The world is replete with all sorts of replicators that team up in whatever ways best allow them to cooperate toward their mutual survival. Teams of replicators can further team up with other teams in a never-ending process of building hierarchical complexity from the ground, up.

It is becoming ever more clear that the grouping of replicators by how they are packaged into phenotypes is of secondary importance to the grouping of replicators by the notion of cooperation, wherever they may exist. Replicators from many different substrates now cooperate very effectively within hierarchical structures that take on characteristics of individuality at their highest levels. For instance, we think of the Internet as a single entity, even though it is composed of many different components spread all around the world.

I am astounded to find that my view of memes is so much more aggressive and unrestricted than views held by many other authors. For example, Dawkins says:

> Memes, unlike genes, don't seem to have clubbed together to build large 'vehicles' – bodies – for their joint housing and survival. Memes rely on other vehicles built by genes (unless, as has been suggested, you count the Internet as a meme vehicle). But memes manipulate the behavior of living bodies no less effectively for that.
>
> – **Richard Dawkins**, *Unweaving the Rainbow* (1998, p.306)

Obviously, Dawkins does not see memes in the same light that I do. From my perspective, memes certainly have clubbed together through various sorts of cooperation to build more complicated memes, and those complicated memes have large phenotypic bodies such as skyscrapers, bridges, cars and airplanes. And yes, I certainly do consider the Internet as a meme vehicle, along with books, magazines, and many other objects having the means for storing and communicating patterns of relationships, which are nothing more than ideas. We'll soon see how even memetic phenotypes themselves – artifacts – can be 'vehicles' for 'housing' the memetic actions that built them.

So, where do language memes fit into this mix? Do they benefit the genes? Well, the real value of language memes, from the selfish perspective of the genes, is their ability to organize many diverse individuals into cooperating groups. Words are able to describe such group behaviors as: the respective duties for members of a family, the mission of a business, the rules of a society, and the terms of an alliance. Languages can facilitate cooperation among many engineers and manufacturers for constructing such complex objects as airplanes, satellites, nuclear power plants, and space shuttles. Whereas genes bring organization at levels from molecules up to the human individual, memes bring organization at higher and higher levels, by organizing hierarchical groupings of individuals.

Language, whether it consists of simple hand motions or elaborate spoken sentences, provides a binding mechanism by which multiple individuals can coordinate their efforts. The synergy that results from many individuals all working toward the same goal gives the group a kind of individuality at a higher level. And all the members of the group become fitter – better able to perpetuate their defining replicators – as a result of their cooperation. I have previously referred to this idea – that cooperative entities form a new entity at a higher level – as the principle of the *inclusive phenotype*.

Only when we view memes as tantamount to the genes of a society can we open our minds to the incredible range of patterns that qualify as memes. And, only then is it obvious that many memes require more than one person to implement them; and hence, without a cooperative society they would never be built. So, not only does a society depend on its many memes, but also, many memes depend on a society for their implementation. Indeed, memes and societies are related in many of the same ways as are genes and humans. A society can even assume a characteristic of individuality – something of a personality – with traits defined by its memetic laws. A society can be kind or cruel, aggressive or complacent. And we'll later see that a society takes on something of an unpredictable mind of its own in its process of collective thinking.

Emergence of Synergy

It seems to be a general rule of nature that complexity begets complexity. Biological evolution has a long history of creating ever greater complexity by building on things of lesser complexity. And, the trend is true of memetic evolution as well. By understanding the complexity of the transistor we were able to invent computers, and once we had computers, we were able to use them to design better, more complex computers. Modern computers have gotten extremely complex by combining far more transistors than ever before. It seems that increased complexity is required for increased cooperation among replicators. It is something of a prerequisite for synergistic cooperation.

Just as rich people tend to get richer, fit replicators tend to get fitter. And, the fitness of replicators tends to correlate with higher complexity. Just as capital equipment can generate more capital (in the form of profits), complexity can generate more complexity, especially through recombination of complex entities. All self-reinforcing trends tend to breed complexity as a by-product of discovering better ways of facilitating cooperation. I find the idea of naturally evolving complexity to be intriguing, and I believe we can bring some clarity to the issue by reconciling it with our new understanding of synergistically cooperating patterns. To do so, we must first try to get a handle on the notion of synergy. So, let us consider a simple example.

Hydrogen and oxygen are both gases at room temperature. But, molecular combination of the two gases produces a liquid – water – having properties completely different from either of the two constituents, considered separately. We may think of the emergent characteristics of water as something of a synergistic product of 'cooperation' between the hydrogen and oxygen atoms. It comes out of nowhere from the inherent properties of cooperating atoms. The emergence of synergy is not supernatural; it is always a very natural, reliable and repeatable consequence of things acting cooperatively.

Synergy doesn't count for much until it participates in the process of pattern replication and natural selection. For instance, it doesn't really matter to anyone or anything whether atoms of hydrogen and atoms of oxygen are in their separate gaseous states or their combined liquid state, unless there is some instance of life nearby that is dying of thirst. In a more general sense, when evolving patterns of molecules, such as genes, are able to produce synergy toward their own perpetuation, they gain an advantage over other replicating molecules having no synergistic benefit. Synergy only matters to the extent that it facilitates life.

Economists have certainly noted the beneficial effects of synergy resulting from the many forms of cooperation among the individuals of a society. Those synergistic effects include great efficiencies from division of labor, specialization, economies of scale, and diffusion of risk. Yet, most of us don't really appreciate what it is that makes a society good. The answer lies in the creation of synergy. But the magic of synergy – getting something from nothing – can only occur through the combining of cooperating elements into something of an individuality at a higher level.

We typically categorize governments as being either good or bad on the basis of how well they protect individual human rights. But from nature's perspective, goodness derives from the fitness of the unified whole, rather than from the fitness of the individual constituents. I gladly admit that the fitness of a unified society depends, to some extent, on fitness at the lower level – the level of individuals – and, for that reason, good governments are those that allow individuals to pursue their own well-being. But, at a more fundamental level, the fittest societies will be those that most effectively foster cooperative synergy among their members, toward the perpetuation of the unified society itself.

Such is the nature of life: Replicating patterns of things establish cooperative partnerships with other replicating patterns, all acting toward their mutual perpetuation. What we see as ever more complexity, growing out of evolutionary processes, is more accurately described as ever more synergistic relationships, requiring growing complexity to make the relationships work. Nature doesn't care for complexity for complexity's sake, just as I don't prefer a lace doily over a simple handkerchief when I need to blow my nose. I prefer whichever works best, and so does nature. Patterns that efficiently gather and direct energy toward their own replication are preferred by nature. It just happens to be a statistical law of nature that the most efficient patterns at replicating tend to be the ones that have found ways to achieve synergy through cooperation with other patterns. And it is the nature of synergy that it generally entails complexity.

Lamarckian Inheritance

So far, the discussion in this chapter has focused on similarities between genes and memes. We are now in a position to highlight what appears to be a very important difference between them. The relevant concept, referred to as *Lamarckism*, actually involves an erroneous belief about evolution that many people used to hold. The belief came from the Chevalier de Lamarck in the early nineteenth century when he proposed an evolutionary mechanism of inheritance. Lamarck can be credited for having advocated a form of evolution well before Darwin proposed his theory on the origin of species. But, unfortunately for Lamarck, his name is now associated with a significant blunder.

Lamarck believed that the process of inheritance would propagate acquired changes in an individual's body to its offspring. This erroneous belief in *the inheritance of acquired characteristics* is a predictable consequence of the common misconception that individuals replicate themselves. Lamarckism was scientifically rejected after the discovery of DNA by Crick and Watson, which clearly showed it is genes that get replicated, not individuals. Indeed, Dawkins has often pointed out that genes are the focus of evolution, not their respective phenotypes.

Consider some examples of Lamarckism: A man who performs laborious tasks, building large muscles in the process, might expect his future male offspring to acquire the characteristic of large muscles. A man who loses a couple of fingers in an accident might expect shorter fingers in his progeny. And a woman who bears many children might expect a propensity for her daughters to be born with wider hips. We now know that the DNA passed from a parent to a child cannot be affected by any sorts of changes to the body of the parent. There is no channel by which information regarding acquired characteristics can get into the natural flow of inherited DNA.

There is a unidirectional arrow of causation pointing from genes to individuals – from the seed to its phenotype. A system is said to be Lamarckian only if it allows *reverse-engineering* from the phenotype back to the seed. A system is said to be non-Lamarckian if, like genetic inheritance, it always employs a unidirectional arrow of causality from the seed to the phenotype.

We might speculate that a Lamarckian system would be in some ways a better system. For example, wouldn't it be nice if the often-performed removal of tonsils and appendices resulted in their ultimate elimination from the human genome? We would then be able to guide evolution's progress with our medical procedures. Even medicines for, say, treating the adverse effects of asthma, would cause them to lessen over the generations. And, women who get breast implants would have more voluptuous, buxom daughters than they otherwise would have had. What a wonderful world it would be … from a male perspective. Please allow me this moment of indiscretion. I promise it will make more sense after several pages.

It seems that evolution agrees: A Lamarckian style of inheritance can be a better system than the unidirectional style that has for so long governed our biology. And so, it appears that evolution has discovered a means for certain acquired characteristics to be propagated from parents to their children in the manner of Lamarckian inheritance. While this newly evolved mechanism does not apply to biological *form*, it absolutely does apply to biological *function*. As important as the structure of the human body is, equally important is the movement of that structure. And acquired patterns of behavior – common movements of the physical structure – can certainly be propagated down through the generations by simple mimicry.

If a man discovers a new technique for catching food, he can teach that technique to his offspring. And thus, an acquired characteristic is 'inherited'. The Lamarckian mechanism enabling culture and technology is a relatively recent evolutionary innovation. Humans are the first species to have perfected the general process of handing down to descendent generations all sorts of beneficial behaviors. As I previously mentioned, vertical memes can be considered almost as virtual extensions to genes, but, quite unlike genes, they are Lamarckian extensions.

Whereas biological inheritance is directional, memes have only a vague sort of directionality. Whereas patterns of genes are in no way discernible from their biological phenotypes, the phenotypes of memes can often be easily reverse-engineered to reveal the patterns of how they were produced. It is this capability of reverse-engineering that allows memes to evolve in a faster and more directed manner. Memetic evolution is a more focused and efficient process of discovery, as opposed to the random discovery of genetic evolution. The greatly improved efficiency of discovery makes the memetic evolutionary process more intelligent.

The ability to reverse-engineer a process allows a human mind to first imagine a goal, and then figure out how to get there. Nature's biological evolution can do nothing of the sort, and is therefore much less intelligent. It is quite premature for me to describe the mental processes by which reverse-engineering occurs, except in the vaguest of terms. At this point, I will present only enough information to give you confidence that such an automatic process exists, and that my later descriptions of cognitive algorithms will account for it. So, very briefly, the human mind achieves much of its remarkable intelligence through the reverse-engineering of memes, and the reverse-engineering process goes something like this:

The human mind automatically learns all sorts of behavior patterns – memes. All learned behavior patterns have certain required antecedent (input) states, and they all produce certain consequent (output) states. As learning takes place, neural connection patterns come to represent the learned behavior patterns. We'll eventually discover that the patterns are automatically sorted by the brain as they are learned on the basis of both their input states and output states. Memetic behavior patterns are remembered and stored by mapping their inputs and outputs to various places in the brain. So, all behaviors requiring a particular input state have their inputs mapped to a particular location in the brain corresponding to that input state. And all behaviors producing a particular output state likewise map their outputs to a location in the brain corresponding to that output state. Now, realize that the notion of 'state' is a very abstract concept, and we are not yet prepared to understand how the mind encodes various states of the world.

When the output state of one behavior maps to the same location in the brain as the input state of another behavior, then the two behaviors can be logically connected. They can form a logical two-step plan. By automatically sorting imagined behaviors in such a manner, the brain is able to quickly find strings of behaviors – plans – that lead from the current state to some goal state. And the search can proceed from the current state forward, or from the goal state backward. So, imagined plans of action can be explored by a search in the forward direction; but, more importantly, plans for getting to a particular goal state can likewise be identified and decomposed into their constituent component behaviors through a backward style of search. It is the backward style of search that yields the capability of reverse-engineering.

We often only have to *see* the phenotype of a meme to get an understanding of how to build it. That is, we are equipped with the mental ability to reverse-engineer memes from their phenotypes back to the activities that produced them. In Dennett's words (1995, p.348): "A wagon with spoked wheels carries not only grain or freight from place to place; it carries the brilliant idea of a wagon with spoked wheels from mind to mind." Now, let us consider how the meme for a wagon can be Lamarckian. If the owner of a wagon beneficially modifies it to some obvious extent, then anyone who sees the new wagon, and is capable of reverse-engineering the modification, can thereby cause the wagon meme to inherit the acquired characteristics of the modification. New wagons are likely to be built in the new style.

For instance, suppose that some owner of a wagon fabricates wheels that are much bigger than everyone else's wagon wheels. When other people see that the big-wheeled wagon can easily negotiate rough and muddy terrain, where everyone else's wagon gets stuck, then the value of a big-wheeled wagon is immediately conveyed to others. They may now imagine themselves riding to the general store without getting stuck. The characteristic of big wheels is then automatically acquired by some descendent copies of the meme for wagons. No wagon builder needs to see a blueprint for the big-wheeled wagon to know how to build it. Since every wagon builder already has all the necessary constituent memes for building wagon wheels, those builders of wagons can easily reverse-engineer, or decompose, the big wheels into combinations of those memes.

The human mental ability to reverse-engineer memetic phenotypes, back to their constituent memetic activities, can also result from a cognitive process of iteratively combining and evaluating various forward-going strings of causal relationships. In other words, we humans have the ability to figure out how modifications were likely accomplished, by imagining what sorts of activities could have produced them. Essentially, a brain asks itself: of the many activities I know how to do, is there any combination of them that will produce the desired result?

As a very rough analogy, consider this problem. You have seven bags of gold pieces containing, respectively, 12, 17, 29, 34, 45, 66 and 87 pieces. You want to exchange some of them for something that costs 96 gold pieces. Is there a combination of bags that adds up to exactly 96 pieces? Aside from a few tricks to narrow down the choices, there isn't much you can do but try various combinations. But, if, as a matter of good fortune, you have already memorized the fact that $17 + 34 + 44 = 95$, then you can easily see the answer to the problem. In a similar manner, if you have a bunch of memes in your brain, and if you happen to know how to combine the memes to construct some useful products, then you will likely be able to reverse-engineer anything *similar* to what you already know how to construct.

So, the process of reverse-engineering can be reduced to an evolutionary process of combining and evaluating various memes in various ways. But, recognize that the brain intelligently confines the evolutionary search to only relevant memes, like sorting the pieces of a jigsaw puzzle on the basis of color or shape before searching for a particular piece. Relevant memes are those having roughly complementary interfaces that, when combined, might stretch from the current state to the goal state. While the search for strings of memes can proceed in both the forward and backward direction, we'll find that detailed cognitive evaluation of resulting strings goes only in the forward direction.

Things get a little trickier when the recipe for constructing some improved feature is very complicated. For example, let's fast-forward from wagons of the past to present day cars having anti-lock brakes. I have only a vague idea of how the anti-lock braking systems work. And so, I would have a great deal of trouble reverse-engineering an instance of it. But a car designer who knows all about speed detection systems and braking systems could probably easily reverse-engineer the anti-lock brakes on any given car. Obviously, the Lamarckian capability is strongly related to the prior having of experience and knowledge, the products of which are mentally encoded as memes.

Some things are much easier to reverse-engineer than other things. For example, the process gets really difficult in the following situation. Suppose, after eating a piece of cake, you are expected to bake a copy of the cake without seeing the recipe. There are some analysis tools that you might use to analyze the ingredients of the cake, but you may need to know information on the sequence of processing the ingredients before you can reconstruct a faithful copy. Someone like Julia Child[4], on the other hand, might have a recipe in her mind for a cake that tastes similar, and she might think to herself, "if I just add a little nutmeg to my recipe it will produce a cake that is identical."

So, some memetic products are not easily reverse-engineered, but most are. In all cases, reverse-engineering is only possible by people who already hold a wealth of relevant memes that can be cognitively combined in various ways to mentally assemble whatever is to be reverse-engineered. It is the ability to be reverse-engineered that allows memes to evolve so quickly, by eliminating the need for random exploration, through random mutations, going down many irrelevant pathways. And, it is the automatic sorting of patterns, by their respective interfaces, that makes the few valuable combinations of them so easily discoverable.

Now here is an interesting question: If genes are actually subsets of memes (defined as patterns that serve as templates for their own replication), then why is it the case that culture and technology are Lamarckian but genes are not? Well, now, realize that there is nothing about patterns themselves that make them able to be reverse-engineered. It is the environment in which hierarchically organized patterns exist that determines whether or not their hierarchical organizations can be decomposed.

Memetic products of technology are Lamarckian (able to be reverse-engineered) only because we humans are in their environments, and because in fact we built the things in the first place. Genetic products of biology will also become Lamarckian when their environments include complete computer mappings from genes to phenotypic effects and powerful mechanisms for manipulating genes (perhaps within ten or twenty years). We will certainly, at some point, understand which genes make some people smart, and we will eventually use that information to make all future people smart. Genes will then be well on their way to being Lamarckian.

Before we leave the subject of Lamarckism, let's have some fun with it. Dawkins has argued vehemently against there being any sort of Larmarckism in the way that biological form evolves. I completely agreed with Dawkins until one day I thought of a mechanism by which biological form can indeed evolve in something of a Lamarckian manner. The Lamarckian mechanism I am considering applies to women who have had artificial breast augmentation. I suggest that such an act can in fact increase the likelihood of their daughters inheriting larger breasts than they otherwise would have inherited.

I imagine Dawkins rolling his eyes and shaking his head in disgust at my suggestion. I wouldn't blame him. The Lamarckian inheritance of artificially acquired large breast size does seem inconceivable, because any information describing larger breasts has no channel by which it can get into a woman's genes, especially those that will be passed on to her daughters. But, consider the following roundabout line of reasoning.

As it turns out, the daughters of an artificially augmented woman will be likely to inherit the genetic propensity for larger breasts from their father, not their mother. How can this happen? Before I explain, I must admit there is a slight hitch in my argument. For the Lamarckian effect to occur, the woman must have had her augmentation performed before she meets the man with whom she will have daughters. But, with this assumption, I can now build my case.

Any future husband of a woman having artificially enhanced breasts is likely to be a man who prefers women with large breasts. And most men who have genes that cause them to *prefer* women with large breasts are also likely to have genes that, when expressed in a female, will *cause them to have* large breasts. Why? Because genes that work well together tend to travel together through the population. And it is common sense that men having a gene causing them to prefer large breasts are more likely to marry women having genes causing them to have large breasts. Once we realize that genes causing men to *prefer* large breasts tend to pair up with genes causing women to *have* large breasts, then everything falls neatly into place. A woman having large breasts, whether artificial or not, is likely to marry a man having genes that *both* cause men to prefer large breasts *and* cause women to have large breasts. So, such a man's daughters will likely inherit a propensity for large breasts from him. Evolution works in mysterious ways (or is it God that works in mysterious ways? ... Well, what's the difference?).

To my knowledge, there has never been a study of this effect, and I can't imagine anyone wasting their time on such an endeavor. But if there ever is such a study, I volunteer to measure and collect the data. Oops, there I go again![5] I cite this silly possible Lamarckian effect only to reinforce some pivotal evolutionary concepts, and to warn the reader that the effects of evolution can be a lot more complicated than they first appear. As famously and repeatedly quoted, Jacques Monod once said "another curious aspect of the theory of evolution is that everybody thinks he understands it!"

For patterns to be Lamarckian, their successful combinations must be able to be reverse-engineered. It is such an ability to be reverse-engineered that facilitates fast memetic evolution. I will sometimes use the term 'Lamarckian' to describe the ability for phenotypic effects to be reverse-engineered, even though it more accurately refers to the inheritance of acquired characteristics by descendant phenotypes.

Ultimately, our task is to understand how it can be the case that neural connections automatically organize themselves in ways that allow our brains to imagine various sorts of behaviors, and, how neural connection patterns can be algorithmically excited and manipulated so as to enable intelligent cognition. The solution turns out to be relatively simple and straightforward. But, before we tackle that interesting problem, we'll spend a few chapters addressing some philosophical issues. We'll talk about the destiny of life, we'll see how learning automatically occurs, we'll define a philosophical model for human behavior, and we'll discount various philosophical challenges to the idea that a mind, complete with consciousness, can automatically emerge from physical matter.

The philosophical implications emerging from my arguments, so far, point to the strong possibility that we humans exist as we do because the human species, or a similar sort of species very much like humans, was statistically destined to exist. Memes, as well as genes, seem to naturally evolve toward patterns that are best able to perpetuate themselves. Without the intervention of free will, either at the level of individuals or at a higher godly level, everything seems to proceed according to strict statistical rules. And so, even if there is some randomness occurring at the quantum level, the evolution of patterns will still migrate toward particular regions that are defined by nature as being statistically more fit. But, if nature defines those regions of fitness, and nature never changes her mind, then there appears to be a statistically defined destiny for life. The next chapter will explore that fascinating topic.

Chapter 4. The Arrow of Evolution

We now take a small break from the methodical examination of memetic patterns, to consider the philosophical implications of what we have so far discovered. There are some profoundly interesting consequences flowing from "nature's incessant compulsion for self-organization."

Consider the concept of Intelligent Design, which has enjoyed a recent re-emergence primarily among religious people. It is full of needless clutter, from the perspective of Ockham's razor. Indeed, I find it preposterous to suggest that some sort of design intelligence, other than evolution, is responsible for having bridged the 'gaps' that remain in the evolutionary evidence of fossils. Evolution itself is an intelligent design process that is quite capable of having bridged those gaps. And the parsimony of science demands we reject superfluous explanations.

Yet, I must fully admit that subscribers to Intelligent Design have something of a strong scientific ally in the *anthropic cosmological principle*, which supports the idea that the fundamental constants of the universe appear to be finely 'tuned' so as to support life. The most crucial of these constants are: the strength of gravity, the ratio of electrical attraction to gravity, the nuclear binding force, the rate of cosmic expansion, and several other scalar quantities. It is now believed by many physicists that if any of the critical universal constants had been even slightly different, the emergence of life would not have been possible.

The stunning coincidences of nature are described by English astronomer Martin Rees in his book *Just Six Numbers: The Deep Forces That Shape the Universe* (2000). He concludes the book's preface with the following statement: "Our emergence and survival depend on very special 'tuning' of the cosmos – a cosmos that may be even vaster than the universe that we can actually see." The special tuning to which he refers effectively poises the entire universe on something of a knife's edge, balancing it on the delicate and precarious edge of sustained chaos – an improbable but necessary precondition for the emergence of life. And the vaster cosmos to which he refers involves the possible existence of many universes, of which ours just happens to have the precisely tuned constants for enabling life. Or, perhaps, universes are capable of replicating themselves in ways that enable them to evolve toward the greater production of intelligent life.

If the universe is indeed completely or even mostly deterministic, and the forces of nature are just such that self-replicating molecular patterns were statistically destined to emerge, then we may logically conclude that our universe *was* 'designed' to generate life. Such a 'design' need not imply a volitional designer, it could have resulted from the same sort of evolutionary process that designed humanity. It is certainly not my intention to advocate a belief in any sort of an anthropomorphic god. I simply want to know what the hell is going on.

Here is what I clearly see, and what this chapter is intended to show: Not only was life destined to emerge, it was also destined to evolve in a particular direction that would take it through several pre-determined stages. There is, in essence, an arrow of directionality to the way all life, everywhere in the universe, will inevitably evolve.

Now, realize, I am speaking in a statistical sense. I am *not* suggesting that all life is destined to produce exact humanity. I *am* suggesting that all life is destined to eventually produce a species that is as moral and as intelligent as humans are. And, of course, the emergence of intelligence naturally leads to technology. Accordingly, just as life on any planet is statistically destined to evolve through several defined stages of development, so is its technology destined to produce things like light bulbs, things like cars, things like sophisticated weapons and things like computers. These are landmark events, springing naturally from life's quest to direct energy toward its own perpetuation, that we should eventually expect to occur to all life, everywhere. All landmark technological events are probabilistically written right into the laws of nature, just waiting to be discovered.

Statistical Destiny in the Advancement of Life

Suppose we could have complete knowledge of all instances of life, everywhere in the universe. Suppose that all the relevant statistical data were neatly compiled into a book of tables, as if we had found God's ledger of life. We could easily discover how advanced we humans are compared to the average, but more importantly, we could learn a lot about the process of evolution by studying the statistical distributions of various stages of advancement across all instances of planetary life. Should we be surprised to find that the data fits a typical asymptotic statistical curve, with lots of primitive life and relatively few cases of highly intelligent life? We shouldn't be surprised at all, because evolution is a probabilistic sort of process, and all life must evolve by way of the very same probabilistic and deterministic laws of nature. Neither should we be surprised to find that life, everywhere, tends to accelerate as it gets going, with each instance of planetary life being destined, eventually, to experience something like a 'knee' in the up-turning curve – a Cambrian explosion of sorts.

As a result of our earlier discussions on evolving patterns, we should now understand exactly why evolution accelerates. The more replicating patterns exist, the more opportunities there are to find subtle mutations of those patterns that are even better patterns. With enough diversity in the replicator pool, a brilliant 'discovery' may be only one or two mutations away. By loose analogy, the more mice are dropped into a maze, the sooner one of them will find the cheese. This principle of acceleration applies to the evolution of all replicators.

There is another benefit that results from the production of many replicators, completely apart from the mere ability to explore more replicator-space. The more styles of replicators exist, the more opportunities there are for cooperative synergies to emerge from various combinations of replicating patterns. It should be obvious that extensive exploration of replicator space is a prerequisite for exploring cooperation space.

These obvious facts raise an interesting question: What sort of characteristic dimensions are we entitled to use for measuring the advancement of life? If we could understand what it is that defines the advancement of life, then we will have discovered what it is that evolution appears to be 'trying' to do.

Given two planets, each having life, what sort of metric best characterizes which instance of planetary life is more advanced? Whatever we may logically use, it will likely apply to life in its aggregate, not to any individual species. Now, recognize that as life evolves and thereby advances, the one inevitable consequence will be an increasing number of different sorts of replicators. So, then, perhaps we can measure the advancement of life by simply counting the total number of unique replicating patterns it has produced. This metric does indeed apply to life in the aggregate, as opposed to any single species. But, is it true that the degree to which our own planet's life is advanced depends more on its diversity of replicators than anything having to do with the human species? I believe so.

While you may think that some other metric, such as the level of intelligence, would be better for measuring the advancement of life, it seems that the total number of unique replicating patterns indeed captures a measure of intelligence within it. If it is not already true, it will very soon be true that most of Earth's modern replicators are not genetic, but memetic. Now, recognize that it certainly takes intelligent life to produce the many sorts of memetic replicators associated with culture and technology.

Fundamentally, I can't think of a better definition for intelligence, at any level, than the following: *Intelligence is the ability to create unique patterns that are good at replicating.* Certainly, Einstein's patterns of thought were unique. And who doesn't recognize the formula $E=mc^2$? The uniqueness and ubiquity of Einstein's ideas mark them (and him) as intelligent. Anyone can generate unique patterns of thought, but they won't get replicated unless they are useful to the perpetuation of other patterns. It should now be obvious that the intelligence of life in the aggregate will strongly correlate with the diversity of all its ubiquitous replicators.

In all instances of life, the evolutionary production of replicating patterns naturally accelerates. The more diversity there is in a set of replicators, the easier it is to get even more diversity by simple mutation. And, again, more diversity yields more opportunity for cooperation, which yields synergy toward perpetuation of the cooperating patterns. So, it seems that the level of advancement with respect to life is directly related to the speed at which evolution is occurring, which in turn depends on the total number of diverse replicators it has already produced. As such, the level of advancement of planetary life is indistinguishable from its aggregated level of intelligence, which in turn depends on the diversity of its replicators.

Memes require humans to get themselves propagated, at least for now. And so, from what we might typically consider to be a backward sort of perspective, perhaps it is the enabling of memes by humans that makes the human species so valuable, from nature's perspective. Perhaps it is the significant cooperation between the multitude of memes and the genes of humans that best represents earthly life's significant degree of advancement. Such cooperation allows the genes of humans and their complicit memes to direct energy more efficiently than any other known set of replicators toward their mutual perpetuation.

Perhaps it is not the diversity of replicators that is important, but rather, the total level of aggregated cooperation among various replicators. By analogy, it is not the number of businesses in the U. S. that generates so much productivity, so much as it is the cooperative relationships between them that leads to efficient production. If indeed the level of advancement for planetary life equates directly to the degree of cooperation among the various instances of that life, then we may logically describe evolution's goal as 'wanting' to facilitate ever-increasing levels of cooperation, which can only happen with ever-increasing replicator diversity as a prerequisite. Indeed, I'll try to show that evolution has a direction in which it endeavors to go, as if it were an arrow, pointing toward the natural emergence of ever greater levels of synergistic cooperation.

Here, then, is the powerful implication of what I am suggesting: If we were able to 'replay the tape' of evolution on Earth, even with vastly different starting conditions, we would eventually end up with life that may be very different in the details of its form, yet very similar in its basic functional characteristics of intelligence and morality. Just as uncertainty at the quantum level yields to stability at the gross level of aggregate matter, so does diversity at the lower levels of evolution yield to stability in the gross stages of life. On the way to evolution's destiny, we should expect to find that all instances of planetary life will pass through some similar stages of evolutionary development. It is a matter of statistical inevitability.

The Evolutionary Stages of All Life

There are several characteristics of life on Earth that we should expect to eventually emerge in the evolution of all life, everywhere in the universe, at least, every place where it is possible for them to emerge. We may identify these inevitable common characteristics by their tendency to enhance fitness in absolutely any environment. There are also some prerequisite dependencies between these characteristics that ensure they will always tend to emerge in roughly the same chronological order as they did here on Earth.

The predictable stages of evolving life, through which the evolution of all life will naturally progress, form an arrow leading from what we typically think of as *evil* toward what we typically consider to be *goodness*. If philosophy and science are ever to be reconciled, it will only happen around a moral good that is consistent with the advancement of life in the aggregate.

If we can indeed identify an arrow of evolution, then that arrow will allow us to think of nature as having a goal for life. We may infer that the arrow points directly at the style of behavior nature ultimately prefers for life to practice. And all species of life must necessarily evolve toward that style of behavior. I suggest that the arrow of life clearly points at the morality of nature. To appreciate the natural progression of the arrow of life, let us start at the very beginning.

At bottom, we should expect all life to initiate from patterns of matter, like our DNA, that automatically replicate themselves by combining ingredients in their respective local environments. Could patterns of energy be candidates for automatic replication? It is highly unlikely that patterns of pure energy could maintain their integrity against the naturally diffusive forces of entropy long enough to support long-term life. The most likely initial replicators then, patterns of matter, will probably originate in some sort of liquid (on "the edge of chaos"), because that phase of matter is the only one that easily allows large stable molecules to be constantly supplied with drifting ingredients that are necessary for replication.

Now, supposing the emergence of auto-replicating DNA-like matter elsewhere in the universe, what characteristics should we then expect to evolve? Well, once replicating patterns of matter emerge, natural selection will always show a statistical preference for whichever of those patterns are able to alter their local environments in ways that allow them to maintain their integrity and replicate more readily. The construction of something like a cell wall from local environmental ingredients is a natural first step in altering the environment toward protecting the all-important replicating patterns against naturally occurring caustic elements. We may assume that some sort of protective layer will inevitably evolve, if it is possible to evolve. I believe it is safe to conjecture that, in any environment, cellular life is always more fit than non-cellular life, all else being equal.

There can be no question that movable life must be statistically fitter than immovable life, all else being equal. The ability to move, in addition to providing extra degrees of freedom, does not prevent the option of staying still. Thus, an ability to move is always fitter than an *in*ability to move. And therefore, all life, everywhere in the universe, will eventually evolve ways of constructing various styles of protective phenotypic bodies capable of motion. I have already mentioned that some bacteria 'swim' from regions that are depleted of molecular ingredients necessary for replication. Perhaps that is the logical first step toward mobility.

In conjunction with developing the ability to move, we should suspect that life will concurrently develop the ability to sense the environment. Only by sensing the environment in some way can an instance of life determine how best to move. The detection of light, through sensors like eyes, is a particularly valuable mechanism for attaining information about the environment. Evidence shows that the evolutionary emergence of something like an eye has occurred in at least twelve different independent instances for various species on Earth. So, we may expect eye-like sensors to eventually emerge in all mobile life, everywhere there is light.

In addition to the need for consuming ingredients of matter for the purpose of cellular replication, thermodynamics tells us that the process of assembling and moving a phenotype must also consume available energy. And so, as gene-like patterns and their phenotypes become bigger and more complex, they must find ways of directing more energy toward the process of their own replication. They must gather photons (heat or light), or eat things that gather photons, or eat things that eat things that gather photons. Thus, nature always prefers patterns of life that efficiently gather energy for use toward their own perpetuation. In all environments, aggregate life will evolve many species capable of moving toward prey and away from predators.

Once life develops the ability to move, it will eventually discover specific movement sequences that aid in replicator perpetuation, including the construction of tools for gathering energy, such as spider webs, and the construction of shelters for protection, such as gopher holes. So, in addition to evolving patterns of matter, there are also patterns of actions and behaviors that will naturally evolve by the very same forces of selection. As I previously mentioned, the actions can either be programmed by the gene-like replicating patterns of matter as automatic responses to certain perceived characteristics of the environment, or they can be mimicked from other instances of life. When patterns of actions are commonly mimicked from one instance of life to another, then there emerges a new style of pattern – a cultural style – that is able to replicate and mutate.

A *specialized* sort of mimicry is evident even at low evolutionary levels of life. Ants, for instance, demonstrate a form of mimicry by following the trails of other ants. An ant following a trail toward food is mimicking the behavior of whatever ant initially found the food and then laid down the trail as it left. Such a specialized form of mimicry has enabled a real-time intelligence to be exhibited by a group of ants, through the simple evolution of those trails toward ever better sources of food. Wandering ants that find better sources of food need only lay down stronger trails in order to facilitate the 'intelligent' evolution of paths to food. Good paths yield to better paths as a result of random discoveries, just as in biological evolution good species yield to better species. It is an intelligent process of evolution in both cases.

Specialized mimicry will eventually emerge in absolutely any environment, because it allows many copies of the replicators that enable the mimicry to benefit from valuable individual discoveries. And such a primitive style of specialized mimicry will inevitably evolve into a more advanced form of *generalized* mimicry. While many sorts of animals engage in specific mimicry, only humans have the ability to completely generalize the process. Humans can often mimic other humans no matter what they do, and this unique ability has led to all sorts of traditions, rituals and behaviors under the broad umbrella of culture. Cultural behaviors have become critical to our human survival. They prescribe our methods of nourishment, hygiene, social interaction, communication, technology, commerce and politics.

Recognize that memetic behaviors are able to evolve much more quickly than genetic behaviors. The cycle of mutation, evaluation and selection for memes of technology, for example, can easily occur many times within a host's lifetime. The cycle for genes necessarily involves at least a generation of hosts. So, given that memes can evolve much more rapidly than genes, adaptively beneficial behaviors are much more likely to be discovered by memes than by genes. Generalized mimicry essentially kicks the process of evolution into a higher gear. As such, it is extremely beneficial to a species of life, and should be expected to evolve in absolutely any environment. But realize that general mimicry can only emerge after specialized mimicry, which can only emerge after mobility.

There is yet another characteristic that we should expect to emerge in all life, given enough time. It is the automatic contemplation of complex actions composed from strings of primitive actions – the ability for predictive planning. Life, everywhere in the universe, will eventually develop minds that are capable of mentally simulating and evaluating plans for doing things that tend to enhance inclusive genetic fitness. Such minds are then capable of proactive behaviors, which are evolutionarily preferred over minds limited to reactive behaviors.

An ability to mentally simulate reality, in what we refer to as *the mind's eye*, along with an ability to evaluate ideas within that virtual environment, will always tend to enhance fitness, in absolutely any environment. The development of intelligent planning is thus a natural step in the process of all evolving planetary life; but it is only likely to occur after life has evolved the means for general mimicry by which instances of life can collect a wide range of primitive actions that can then be strung together into various plans.

The product of intelligence is technology, and the arrow of evolution quite naturally extends itself well into the memetic domain of technology. Just as we should expect to find that all instances of planetary life, everywhere in the universe, will evolve through certain common stages of 'biological' development, so should we expect to see certain inevitable stages in the evolution of technology. We should expect, for instance, that, because a wheel provides similar benefit to the perpetuation of life in absolutely any planetary environment having land, all intelligent life, once it has achieved the proper mobility for moving onto land, will eventually discover the wheel. The value of a wheel for augmenting mobility is written right into the laws of nature. It is not a matter of human creativity or ingenuity, unless those cognitive characteristics are defined as mechanisms of simple, iterative searching.

It is easy to see that all aspects of technology are written right into the laws of nature, just waiting to be discovered. And many of those discoveries, having natural prerequisites, are destined to occur in roughly the same order for every instance of planetary life. For instance, we shouldn't expect life to invent a wagon before it first discovers the wheel. And the invention of a car will always come after the invention of a wagon.

The inevitable discovery of electricity opens up a flood of innovative possibilities for directing energy toward the perpetuation of life. Those inevitable discoveries include motors, engines, electronic controls, intelligent weapons and computers. All those possibilities will have been hiding in the laws of nature, waiting to be discovered through a natural sequence built on prerequisites. If we want to fully understand intelligence, we must first adopt a view of all human technological achievements as having been statistically destined to occur. Only then can we clearly see that the statistical destiny of all life is determined by the physical laws of nature. What we think of as human creativity amounts to nothing more than statistically inevitable discovery.

All instances of planetary life will inevitably discover language, because language has the potential for enhancing fitness in absolutely any environment. The trails of ants are a form of communication, and even bees are able to communicate by telling each other of flight paths to food. In both cases, an evolutionary process emerges from differential probing (by individuals) and selection (by communication among them). A rich language has the effect of coordinating and synchronizing many individual minds into a single societal mind. And, it will become perfectly clear as we proceed that, like all minds, the mind of a cooperating society operates by a process best described as evolutionary.

The resulting cooperation among many individuals, made possible through language, allows the super-individual – the society – to build things way beyond the capabilities of any individual. For all instances of life, natural language will eventually evolve to include the sub-languages of mathematics and logic, because the operations of those symbolic sub-languages have counterparts built right into the laws of nature.

Once a planetary instance of life has an expressively rich form of language, its succeeding generations can then 'stand on the shoulders' of previous generations in the development of sophisticated technological innovations. The inevitable 'knee' in the accelerating curve then puts technology on its skyward path.

The Morality of Nature

Finally, there is at least one more natural characteristic that will eventually evolve in all life across the universe, after the emergence of mobility, mimicry, intelligence and technology. It is *morality*. Just as mobility, mimicry and intelligence make life statistically fitter in absolutely any environment, so does morality, which fundamentally encourages cooperation among multiple individuals toward the mutual perpetuation of their defining replicators.

As I have said before, and will continue to reiterate, the synergy resulting from cooperation among replicating patterns is the natural essence of all life. In all instances of planetary life, the valuable principle of cooperation is bound to eventually reveal itself at the level of phenotypic hosts helping each other toward the mutual perpetuation of their defining replicators. When it does, those moral individuals will prosper.

Fundamentally, life can be defined as perpetuated synergy. Indeed, perpetuated synergy is evident at all levels of life. As I previously mentioned, molecules of enzymes and molecules of DNA synergistically cooperate to replicate the patterns of both types of those very same molecules. Replicating patterns of various genes synergistically cooperate to construct cells. Collections of cells synergistically cooperate to produce organs of various sorts, and collections of organs cooperate to produce individuals. The most amazing organ, the human mind, results from collections of synergistically cooperating neurons. What we think of as a soul is merely the synergy produced through that cooperation. The moment the neurons stop cooperating, the 'soul' evaporates into nothingness from whence it came.

Nature selects individuals who synergistically cooperate in colonies, hives, flocks, packs, herds, businesses or societies. Even businesses form synergistic partnerships, and societies form synergistic alliances, both of which tend to make the participants more survivable than they otherwise would be on their own. The synergy resulting from cooperation gives the responsible replicating patterns an evolutionary edge over other, similar, non-cooperating patterns in the domain of fitness. And so, a cooperative morality will eventually emerge in all life, especially intelligent life, independent of the environment. A generalized ability to cooperate may in fact require a certain level of intelligence in order to discriminate between cooperators and non-cooperators, and to assess the probability of future benefit that might result from the necessary sacrifices or investments associated with cooperative behaviors.

It now appears that life, everywhere in the universe, will always tend to become more cooperatively moral as it evolves, over the long term. I personally do not seek to ascribe a godly influence behind nature's preference for moral behavior, but I see nothing wrong with so doing, especially if it encourages religionists to accept the truth of evolution. You might notice that I am trying to cooperate with others of differing opinions. It is the moral thing to do.

It is a moral truth, independent of the naturalistic fallacy[6], that, if we want our descendants to enjoy prosperity, we should foster among ourselves ever greater mobility, sources of energy, educated mimicry, intelligence and moral cooperation. Thus, we must embrace and advocate all the things that nature ultimately prefers we practice. If we don't, our genetic and memetic patterns will eventually be replaced by other patterns that *do* cause their hosts to embrace the preferences of nature. Such an evolutionary scenario, causing the eventual replacement of our genes and memes, cannot be good for our children and their descendants.

It should now be obvious that there are clear evolutionary reasons for being loving spouses and nurturing parents, for being friendly neighbors, for being patriotic citizens, and for dignifying anyone who cooperatively contributes to the welfare of the society whether they are white, black, yellow, tall, short, gay, straight, crippled, or otherwise. So long as one can cooperate, one can contribute. The extent to which one does contribute to the furtherance of life determines one's evolutionary worth.

Human cooperation always requires the building of trust, which can only happen among people who are reliable and honest. Indeed, we intuitively believe it is moral to be trustworthy, reliable and honest. Yet our moral intuitions are not generally aware of precisely why it is good to be trustworthy. Now we know: Trust facilitates cooperation, and cooperation facilitates the survival and perpetuation of life. We intuitively believe it is morally good to help others in need. Now we know why: Their feelings of gratitude will likely impel them to help us when we're in need. Reciprocated acts of kindness constitute a form of cooperation that tends to mutually benefit both parties.

The author Robert Wright, in his book *Nonzero: The Logic of Human Destiny* (2000), touches on the principles of what I refer to as *cooperationism*. Wright correctly identifies the roots of human morality, and its inevitable destiny, as springing from the non-zero-sum outcomes of certain types of relationships. He is referring to the fact that cooperative relationships yield a total benefit to the participants that exceeds the sum of their cooperative efforts. So, the product of cooperation, minus the effort involved, is nonzero. I would have preferred for him to use the more common term *synergy* in place of the unfamiliar and somewhat convoluted term *non-zero-sumness*. But, I am very supportive of his reasoning regarding the logical basis for morality, as it is quite in agreement with the principles I express in *The Laughing Genes*.

If there is a style of morality that is preferred by nature, then we humans must be closer to it than any other form of life on Earth, simply because we are more highly evolved. So, if we like the morality that humans practice better than we like the morality practiced by animals, then we should really like the morality to which nature gives ultimate preference. Our intellect now allows us to deduce precisely what it is that nature views as moral. Only through intellectual reasoning can we discover the natural moral truths toward which our genetically defined emotions of love, friendship and compassion vaguely point, and toward which all life will inevitably evolve.

And thus, we have found evolution's primary use for intelligence. It may have initially evolved as a means for increasing the fitness of individuals, but its scope has broadened significantly. It is now meant to facilitate the perpetuation of life through synergistic cooperation among groups of individuals. The primary adaptive value of intelligence today is its ability to predict the likely future synergistic return on effort (or capital) invested into cooperative relationships.

Evolution's Inevitable Future

Evidence seems to suggest that we individuals are autonomous meme processing machines, and as such, we are simply not the volitional authors of our own creativity. Instead, creativity appears to emerge as knowledge automatically evolves. As our brains automatically combine ideas to form more powerful ideas, it is that act of combining that we refer to as creativity. And, as we humans collect more and more valid memes, the number of possible combinations of memes goes up geometrically. Through the development of computers, computer networks, and huge searchable databases, we happen to be creating an environment conducive to the evolution of knowledge – an 'ooze' rich with 'nutrients' and 'enzymes' for knowledge development – and we are seeing memetic evolution progressing on an exponential scale.

Computers hold the promise of being much better meme-processors than humans. Computers are not limited in size or capacity in the manner that human brains are constrained in size by the width of the human birth canal. Note that some memes for constructing computerized robots are currently being phenotypically expressed *by* computerized robotics and are, therefore, close to becoming self-replicating. I can easily imagine a new computerized life form emerging and growing dramatically over the next century. Most of us won't think of it as life for quite a while, but it will evolve, and it will do so at a much faster rate than our biological genes evolve, due to the Lamarckian style of intelligently directed meme mutations. We should prepare ourselves for amazing technological advances coming more and more from computerized processes. The technological advancement of memetic life is already accelerating at an incredible and frightening pace. Dawkins refers to the memetic style of replicator as follows:

> It is still in its infancy, still drifting clumsily about in its primeval soup, but already it is achieving evolutionary change at a rate that leaves the old gene panting far behind.

> – **Richard Dawkins**, *The Selfish Gene* (1976a, p. 192)

Dawkins wrote those words as IBM was preparing to introduce its very first personal computer. In just the thirty years since, computers have become a million times more powerful, for approximately the same value of cost. The pace of *personal computer* evolution is so frenetic that IBM has decided to stop competing in that market.

In recognition of the fact that the advancement of both life and intelligence requires them to be poised on the edge of chaos, the confusion accompanying rapid change is nothing to fear – it should be used to advantage. As the business guru Tom Peters has suggested, the businesses that will succeed in the future are those that are able to find a way of "Thriving on Chaos." But, while we must learn to embrace and manage the rapid evolution of technology, we must also keep in mind that it will inevitably bring some unwelcome changes and some uncomfortable realizations about ourselves. Even the near future will likely be wildly unpredictable. But, the far future … well, anything goes.

Nature's Way

We humans tend to believe that we are responsible for guiding the destiny of earthly life. After all, we are polluting the environment, consuming precious energy resources, destroying the ozone layer, and contributing to global warming. But the thinking entity referred to as 'we' is something quite different from any individual. Humanity, considered in aggregate, might in fact have a mind of its own.

We as individuals may try to influence the mind of humanity, because we think it is important. Indeed, I think there is nothing more important, and that is why I publish the books I write. But whether or not I am successful in promoting my ideas is not determined by me so much as it is determined by the forces of nature that have thus far shaped humanity's collective mind through statistically evolving genes and memes.

The mind of humanity will proceed to evolve in a direction that is absolutely unalterable by any individual. But that doesn't mean that an individual should stop trying to influence the process, because progress requires individuals to move it along. It does mean, however, that an individual can only be effective in influencing its society's thinking by 'pushing' it in a direction that it is ultimately destined to go. An effective individual is one that 'pushes' in the direction that evolution's arrow is pointing. Perhaps I can best illustrate this by returning to the issue of morality.

If we accept that our human morality has been historically defined by the natural and statistically deterministic evolution of our defining replicators, then we are admitting that no individual in the past was responsible for establishing our morality, only discovering it. So, can any individual possibly alter the *future* of human morality? Yes, but only by proposing a style of morality that is even more consistent with nature's laws than the style of morality we already practice. I believe there is such a style of morality based on the principles of cooperationism, and I believe that human morality will eventually move in that direction no matter what any individual says or does. All one can do is help it along by 'pushing' in the same direction. When philosophers go in search of 'moral truths', they are implicitly admitting that we humans don't define proper morality, some higher power does. From my pantheistic viewpoint, the higher power is indistinguishable from nature.

We can more easily see the validity of this concept by looking backward in time, to a prehistoric era when the evolutionary development of the opposable thumb and the human ability for general mimicry enabled the production and general use of primitive weapons. Suppose the *first* hominid to have ever held a spear had decided to lay it down and become a pacifist. Would that event have changed the character of all subsequent morality? Would he have altered the direction of moral evolution such that all the intervening wars, between then and now, would have never occurred? Or, would he have simply been slaughtered by the *second* hominid to have ever held a spear?

A million years ago, hominids were incapable of peacefully conducting their affairs. They just weren't yet memetically prepared for a civilized morality. They had only a primitive sort of language, unsuitable for discussing philosophical issues of morality. They knew nothing of democracy, or commerce, or contractual law. If any such hominid had wanted to promote pacifism, he had first better invent a sophisticated language, then invent an entire system of laws such as we have today. But even then, he would have faced the monumental task of teaching the language to all his peers and convincing them to adopt his laws. Evolution must follow its own course over due time. It is inevitable that the evolution of all planetary life will eventually pass through a state somewhat similar to the state of our world today.

We typically think of morality as a style of behavior that individuals choose, or don't choose, to practice. We haughtily believe that humans choose morality in defiance of the "red in tooth and claw" Mother Nature, as if we humans are superior and more enlightened than 'she' is. Scientific evidence shows our arrogant view to be backwards. Morality is more accurately viewed as a style of behavior that nature prefers we practice; the choosing is statistically performed by natural selection, not by humans. Nature reveals her preferences by selecting those genes and memes that inspire the behaviors we have evolved to think of as moral.

From a gene's-eye-view, it is scientifically obvious that we humans care about the welfare of our children, not because we *choose* to be moral, but because we are *programmed by our genes* to care. Such caring is the first and most fundamental aspect of morality that nature eventually breeds into all life. It forms the first rational justification of sacrifice by one individual for another. Genes that inspire parental sacrifices toward nurturing their young simply prosper as a result of those sacrifices. It now seems clear that human genes cause human parents to *want* to sacrifice for their children. Parental love is certainly responsible for most of the sacrificing performed on Earth, but we must suppose it is, in all cases, inspired by natural evolutionary forces operating on the human replicators of genes and memes.

Aggregate life can't ultimately evolve in any other direction than what is approved by nature. Through its inexorable probing, evolution will eventually discover nature's way. And prosperity goes to whatever segment of life happens to discover it first.

Chapter 5. **The Automatic Nature of Memes**

As clearly as we humans are products of evolution, so is it also clear that human minds must have been evolutionarily designed to deterministically pursue whatever courses of action tend to perpetuate human genes. Nature would surely have selected against any sort of mind having a freedom of will capable of whimsically choosing alternative acts that are detrimental to the propagation of its defining genes. Some human minds *do* choose such acts in the commission of suicide, but logic demands that we conclude those minds to be defective in their programming, either through their genes or memes, not that they have free will.

Even though strongly counter-intuitive, all relevant scientific evidence suggests that the flow of human thought is mostly deterministic. Indeed, brain researchers now understand in significant detail the electro-chemical properties of neurons that allow them to act as they do. They know how certain ions get pumped through the cell wall to build up a reservoir of electric potential for the spike that is to propagate down the axon. They understand how neurons connect to other neurons through synapses, and they understand the basic role of neurotransmitters. No longer should we have to debate whether the mind is mostly if not completely deterministic. It is scientifically obvious. One might say it is a 'no-brainer'.

If human minds are deterministic, then there must be an automatic mechanism by which memes are able to get themselves propagated from mind to mind to mind. In fact, the mechanism is quite straightforward and easy to understand, at least, from a high-level perspective. I believe the simple mechanism has been overlooked for so long simply because there are so very few scientists who have completely accepted the obvious truth about the deterministic nature of the human mind, and who are willing to admit it.

This chapter is intended to show how the evolutionary development of emotions laid the groundwork for the automatic propagation of memes. There need not be any sort of volition nor spiritual intervention nor freedom of will to explain our human characteristics. We have all the information we need to understand how the brain functions. I'll even show in the next chapter how consciousness can be reduced to something of a computational trick. And I'll explain exactly why our genes found it adaptive to play such a trick on us.

Memes propagate, very simply, by triggering our automatic emotions. But before I make that case, I want to start-off the discussion from the viewpoint of an outdated model of automatic behavior that became popular as a result of B. F. Skinner's work.

Behavorism

Harvard psychologist B. F. Skinner has described experiments in which an animal establishes a cause-and-effect relationship between some action that it performs and a reward that is given to it in response to that action. In what is called a 'Skinner box', a pigeon is given a piece of food whenever it pecks at a certain button. The pigeon eventually *learns* how to get more food by pecking at the button. The experiments give us reason to believe that some primitive components of high-level cognition, such as learning, are common to lower forms of life, as we should certainly expect from evolution's incremental style of continuous improvement. But, on the other hand, we are incorrectly led to believe that some sort of reward is an essential ingredient to learning. As we shall soon see, it is not.

If pigeons can think on some rudimentary level – that is, if their actions are intelligently guided by relationships learned in the past – then we have some evidence that the capability for intelligence evolves. Lower forms of life, like pigeons, have a little bit of it, and we humans have a lot of it. The interesting question then becomes: What are the characteristics that differentiate human intelligence from pigeon intelligence, or from dog intelligence, or from monkey intelligence? The answer, I am inclined to believe, has to do with the human ability to develop and assess long and complicated plans.

While animals live 'in the moment' of the present, pursuing immediate gratification, we humans are willing to sacrifice now, for greater rewards in the future. But how are we able to determine which courses of action will yield future rewards? The answer represents the essence of intelligence, and is really quite simple. We mentally collect and remember various patterns of behavior by observing the behaviors of others. In addition to remembering the behaviors themselves, we also have the incredible capability of remembering the causal effects of those behaviors, relative to the environment. That capability allows us to predict the effects of various behaviors, or strings of behaviors, in the context of the current environment. By that mechanism, we are able to construct plans in order to achieve certain goals. Let us consider this mechanism of contingent mimicry in more detail.

It is convenient to talk about our cultural behaviors as if they are linear scripts – recipes for living. Over our lives, we accumulate many scripts just by watching others. We have a script for every occasion. For instance, when eating at a restaurant, we expect a waiter or waitress to bring menus, take orders, bring food, and finally, when the meal is over, bring the check. We aren't born with our scripts; we acquire them through observation. We learn proper behaviors through mimicry of others, especially our parents.

A proper conceptualization of these behavioral scripts requires us to view them as a web of possible activities that we might undertake. In this web, various scripts intersect at various points, similar to the words of a sparse crossword puzzle. For example, the act of brushing my teeth is in both my script for going to bed at night and my script for getting up in the morning. Whatever part of my brain is responsible for how I brush my teeth, it must be connected to at least two different scripts. And, the act of brushing my teeth is itself a script that entails grabbing a toothbrush, some toothpaste, applying the paste to the brush, inserting the brush into the mouth, etc. We have scripts for everything that we routinely or even occasionally do, and they are all interconnected by a plethora of pathways. The intersecting pathways form a complicated web, connecting many nodes, each representing either a sensed event or a possible activity.

It is somewhat difficult to describe how such a connection web might work, but it is worth wading through several paragraphs to visualize a very high-level and simplistic sort of model. The model will represent a neural architecture that became popular as a result of Skinner's behavioral experiments. We'll later make a slight but very significant refinement to the model so that it is consistent with a memetic perspective. We will be able to easily identify the structure of a meme in that refined model.

Let us start by imagining that every possible activity is represented by a light bulb in a huge web of interconnected light bulbs. When a bulb representing some activity is brightly lit, the corresponding activity is being considered for performance. In addition, let's imagine a bunch of input perceptors represented as light bulbs in this connected web, as well. Various perceptor bulbs light up as a result of various events being detected by the senses. Now let us get more specific in the contemplation of an example.

Among the many perceptor bulbs, there is one that senses when the alarm clock goes off in the morning. There is also a bulb that corresponds to the activity of rising out of bed, and yet another that corresponds to hitting the snooze button on the alarm clock and going back to sleep. Let us imagine that there can be connections between the various bulbs, allowing energy to flow from a bulb that is brightly lit to other connected bulbs. Various connection pathways can have various strengths, some conducting much energy, and others allowing little or no energy to flow. We'll postpone the discussion of how these connection pathways change over time in the process of learning. Let us just assume for now that they have become properly organized as a result of learning.

Suppose there are connections going from the *alarm* bulb to each of the *rise* and *snooze* bulbs, such that, when the alarm goes off, the brain automatically contemplates rising out of bed or hitting the snooze button. Now, let's presume that the connection to the *snooze* bulb is stronger than the connection to the *rise* bulb. When the alarm goes off, it causes the corresponding *alarm* perceptor bulb to light, and energy to flow to the *rise* and the *snooze* bulbs, of which the *snooze* bulb is the more strongly connected. Since the *snooze* bulb receives more energy and is more brightly lit, one hits the snooze button and falls back to sleep. Our simplistic model is such that activity is determined by the brightest bulb.

Now, let's suppose that the alarm goes off again after ten minutes. Again, the *alarm* perceptor lights its bulb, and once again, the *rise* and *snooze* bulbs receive energy. But this time there is another perceptor bulb that has become lit; it is the perceptor that recognizes the situation in which the alarm has gone off twice. It makes sense to suppose there is a strong connection between this *double alarm* perceptor and the *rise* bulb, which causes the *rise* bulb to glow brighter than the *snooze* bulb. Thus, one is forced by the circuitry in one's brain to rise out of bed, all the while, feeling as though one has volitionally chosen to do so.

On some occasion, when there is an important project going on at work, there may be a perceptor in the brain that glows in recognition of this important event, and there may be a connection between this perceptor of importance and the *rise* bulb, a connection that causes the *rise* bulb to glow brighter than the *snooze* bulb on the first occurrence of the alarm. And we might also speculate that there is a connection between the perceptor that senses pressure in the bladder and the *rise* bulb. Such a full-bladder perceptor being active may also cause one to get out of bed on the first alarm, depending on the strength of the bladder pressure, and hence, the brightness of the corresponding perceptor bulb.

What about those times when we are in quiescent conditions and have nothing to which we need respond? We can visualize all the bulbs in the vast array as being slightly lit at all times, the brightest of which determines the focus of our conscious thoughts and hence our next action. Even when there is no pressing need to respond to anything – for instance, after we've relaxed in a quiet environment and all our perceptor bulbs have gone dim – there is still one bulb in the web that is the brightest. While the brightest may not be very bright, it still captures one's attention. Perhaps the brightest bulb turns out to be the one that recognizes the important situation at work, and consequently, one's conscious thoughts are directed toward that area of thinking. Or, perhaps the brightest bulb is one that lights as a result of some habit, such as smoking. Or, perhaps it is the bulb that senses hunger, causing one to embark on a journey to the kitchen.

There is a bulb for everything that we like to do. The bulb for watching television may glow brightly for many people, causing them to choose that action in the absence of more pressing, required activities. There is a bulb for going to the movies, for going to the park, for taking a walk, and so on. Many of our recreational bulbs involving the outdoors are strongly connected to the weather perceptor bulb, which only glows when the weather is nice. When the *nice-weather* bulb glows brightly, it allows energy to flow to all the bulbs corresponding to outdoor activities.

Now we need to relate this model of connected bulbs back to our previous view of scripts. We can certainly establish connections from bulb to bulb in a manner that forms a linear script; however, when we get to an intersection between two scripts, we need the ability to continue on with the script that got us there. For example, my *wake-up* script and my *go-to-bed* script both intersect at the activity of brushing my teeth. When I finish brushing my teeth, how does my brain know which script to follow? That is easily explained by perceptors in my brain that detect morning and night, and that those perceptors are properly connected to the steps following teeth-brushing in both of the intersecting scripts.

Also, each step in a script can be dependent on connections from bulbs much earlier in the script. Perceptors can then recognize when much earlier steps in the script have been completed. So, as the brain follows a particular script from bulb to bulb, there is some energy sent to the bulbs corresponding to later steps in that script, thereby preparing them, or sensitizing them. If the current script intersects with another script at a particular bulb, the current script is properly resumed on the other side by virtue of the previously sent energy to later steps in the script. In other words, the act of eating breakfast sensitizes the bulb corresponding to driving to work, even though full activation of that bulb requires additional energy from either the act of brushing my teeth, which is the next step in the script, or from the perception of being late.

The behaviorist network I've been describing explains behaviors that can be potentially quite complicated, yet completely deterministic and totally responsive to external and internal perceptions. I find the model to be somewhat consistent with my history of behaviors. For example, I often make mistakes while driving, turning the wrong way at an intersection as if I'm dropping my kids off at school, while my original intent was to go the other way toward the store. I am a creature of habit, and when I reach an intersection in the roads, I am also reaching an intersection in the scripts within my brain. If my conscious thoughts are engaged in some other endeavor, then my subconscious mind can easily lose track of my complicated network of scripts, especially at points of intersection.

For this connectionist system to be an accurate model of an intelligent brain, the connection strengths between various bulbs must be adjusted through some process of learning. But how would such learning take place? The traditional approach goes something like this: The system always automatically arrives at *some* decision regarding what to do next. Before learning, the decision is random, so, depending on whatever bulb just happens to be brightest, that bulb's corresponding activity is chosen for execution. After the activity is performed, the system either experiences a good feeling, a bad feeling or something in between. For instance, if the activity decided upon was to pluck a berry from a bush and eat it, the resulting feeling would depend on how the berry tasted. If the berry tasted sweet, then the corresponding good feeling would automatically cause any recently active connections to become strengthened. If the berry tasted bitter, then recently active connections would become weakened or even inhibitory.

This sort of *connectionist* network has been, for a long time, a very popular model of the brain. And, indeed, it has inspired much research into various architectures of *neural networks*. Researchers have developed techniques, such as the *back-propagation* algorithm, for distributing 'reward' backward through the system as a means for strengthening whatever connections were responsible for producing an outcome deemed good enough to be rewarded. Neural networks have indeed achieved the ability to learn how to perform a limited class of functions on a limited class of input patterns. But they seem to have reached a plateau in their abilities, way short of human cognitive ability. We'll soon find the mechanism of reinforcement by reward to be unsuitable as a basis for human learning.

Learning by Reward or Punishment

The idea of strengthening connections on the basis of reward is a logical one, given all the behavioral data provided by Skinner. But, while such a learning process may work quite well for systems that live 'in the moment', like most animal brains, it fails miserably for intelligent brains capable of imagining the future and constructing elaborate plans. Yet it is the predominant method for training most artificial learning systems, such as specially programmed computer software and neural networks.

As I previously mentioned, the commonly accepted learning technique for neural networks is known as the *back-propagation* algorithm. The networks are typically trained by presenting some input pattern – a learning stimulus – then looking at the output to see how well it performed. The difference between what the output is and what it should be can be thought of as either penalty or reward, depending on whether the outcome is bad or good. The reward/penalty currency is then 'given' to the output node to be distributed to all the connected nodes that recently fed it energy and caused it to fire. Those nodes then turn around and disperse their newly acquired reward/penalty currency over whatever connected nodes caused them to fire.

The currency propagates from the output, backward through the system, strengthening or weakening connections as it goes. The backward propagated currency is intended to reward whatever connections pushed the output in the right direction and penalize whatever connections pushed the output in the wrong direction. But many connections that were only incidentally involved in the outcome will be affected as well. Now let us consider this learning mechanism from a memetic perspective.

It is very difficult to imagine how a style of learning based on the back-propagation of reward could possibly be applied to the encouragement of mimicry. There is rarely any immediate reward resulting from a mimicked behavior. Indeed, the kind of behaviorism that Skinner made popular with his experiments on animals may yet be improperly influencing our attitudes toward human learning, by diverting our attention from the more important mechanism. While we certainly see the effects of positive reinforcement in animals and even in humans, we must guard against making the unfounded assumption that *all* human learning is guided simply by reward and punishment. Such a misconception is a natural consequence of the fact that reward-based behaviorism was, for a long time, the only recognized basis available for the automation of behavior.

Marvin Minsky saw the problem long ago. Commonly recognized as the 'father' of artificial intelligence, Minsky was the original proponent of a mind model involving many different modules acting independently. His revolutionary book *Society of Mind* (1985) laid the groundwork for such a model. In that book, Minsky considered how the reinforcing aspects of reward might be distributed over whatever modules are responsible for a given behavior. He rightly concluded (p.76) that: "We cannot learn to solve hard problems by indiscriminately reinforcing agents or their connections."

The problem with using reward as a basis for learning becomes apparent when choosing a time frame over which to distribute the reward. By strengthening whatever connections were active just before a reward was achieved, the behavior associated with those recently-active connections will indeed become more likely under the same environmental conditions in the future. But, what do we mean by 'recent'? If we strengthen the connections that were active during the few seconds before a reward was achieved, then we cannot expect to learn plans that take more than a few seconds to perform. If, on the other hand, we strengthen the connections that were active during the hour before a reward was achieved, then we will likely include many behaviors that had nothing to do with the achievement of reward. And we are still limited to learning only plans of less than an hour's duration.

So, while Skinner's style of behaviorism has properly fallen out of favor as a means for explaining human intelligence, I believe that some of the residual implications still pervade much of the thinking in that particular area. A belief in the need for some sort of reward still haunts our thinking on the nature of learning. And, consequently, the notion of reward remains common in artificial learning systems, even though smart people are figuring out ways to avoid the problems associated with the time frames. Let us consider a particularly interesting technique.

One of the first computer programs capable of learning, perhaps the most famous, was written in the late 1950's by an IBM computer designer named Arthur L. Samuel. The program was able to *learn* new tactics for the game of checkers. Samuel worked, at the time, in collaboration with John Holland, who has gone on to develop many programs that are designed to learn. In his book *Hidden Order* (1995), Holland describes his mechanism of distributing reward to sub-components that are likely to have been responsible for producing good results. As a side note of interest, he also describes a process similar to meiosis for mutating and re-combining the sub-components of behaviors in ways that allow his programs to evolve.

Holland's technique for distributing reward has no time frame. It is more like a free-market system of commerce, with reward acting like money. When an output node is active during times of reward, it gets a sort of currency that it can use to strengthen its connections with any feeder nodes that may have excited it. However, the currency is not passed to the feeder nodes that excite it at the time of the reward, but instead, is saved and given to feeder nodes that excite it *in the future*. Connections between nodes are strengthened whenever currency is exchanged. Lower nodes that receive currency then save it for strengthening connections with *future* suppliers of activity below them. Notice that the currency doesn't necessarily go to the nodes that caused the good behavior, which then resulted in reward. Yet, Holland's systems do exhibit learning. The theoretical basis for such a mechanism of learning rests on the assumption that connections consistently leading to reward will statistically get properly strengthened over time. While this idea is interesting, perhaps we can do better.

There clearly is some connection between learning and reward. But, we have been led down the garden path of believing that learning requires outside evaluation and reward, like a teacher who goes around the class, checking the work of students and giving candy to those who perform correctly. There is a completely different mechanism for learning that doesn't involve any sort of reward whatsoever. It is a mechanism that learns all the time, without any sort of outside evaluation of outcomes, and without reward. Evidence suggests this automatic mechanism of learning is responsible for most, if not all, of the learning that goes on in the human brain. But to understand it, we first need to modify the architecture of the connectionist network.

Hierarchical Connectionism

It is difficult to imagine how a web-style of connectionist network would map to the encoding of memes. It is tempting to view the connection patterns as corresponding in some way to various memes, but where would one meme end and another begin? Recall Kate Distin's suggestion that memes must have a particulate structure, with clearly defined boundaries, if they are to be meiotically shuffled and recombined. It is very hard to partition the jumbled view of a connection web into nuggets of mimicked behaviors or individual concepts. We should be very suspicious of the previously described architecture of connectionism because it just doesn't support the sort of intelligence that we now believe to be adaptive – it doesn't explain automatic mimicry. Allow me to describe a slightly different model that will map directly to a memetic sort of intelligence.

The flat, complicated network I recently described can be better understood by adding a new vertical dimension to it. Let us treat the perceptor bulbs as being at a lower level than all other bulbs. They receive sensory input from the environment, both external and internal to the body. They are on the front lines, sort of like salespeople for an organization. We may consider them as being at the lowest level. Now, let us elevate all the rest of the bulbs to higher levels depending on how far removed they are from the perceptor bulbs. Bulbs that are directly connected to perceptor bulbs form the next higher level, while bulbs that are far-removed from all perceptors are at the highest levels. What we get is a bunch of overlapping hierarchies. Perceptors are at the bottom of every hierarchy, like the tips of the roots of a tree, and they feed their perceived energy upward to higher and higher levels, eventually reaching some top-level node that is representative of an abstract concept.

The hierarchies overlap in the same manner that two root structures will overlap if they are connected to trees planted right next to each other. Since perceptors, such as the photoreceptors in an eye, must be shared among all the visual hierarchies in the brain, we may visualize multiple overlapping hierarchies as something like closely planted trees whose root tips all meet up at various places deep in the soil. Neural hierarchies have their root tips meet up at sensory elements, such as the many individual photoreceptors in the retina of an eye.

Hierarchical structures have long been recognized as very powerful tools for describing all forms of complexity, including all the relationships inherent to life. Indeed, in an article titled "The Architecture of Complexity" (1962), Herbert Simon taught us a lesson that, in Dawkins' words, suggests "a general functional reason why complex organization of any kind, biological or artificial, tends to be organized into nested hierarchies of repeated subunits" (1982, p.251).

We shouldn't have expected any differently, given our understanding of how synergistic cooperation forms the rationale for every stage of evolutionary advancement, no matter how small and no matter what sort of replicators are involved. Cooperation between elements at any level binds those cooperating elements into a logical entity at a higher level – this is the principle I refer to as the *inclusive phenotype*. It is true for the cells of an organism; it is true for the members of a society; it is true for all employees of a business, and it is true for components of technology. Since logical entities can cooperate and thereby form another entity at an even higher level, the result is always a hierarchy of cooperative relationships.

It appears that hierarchies are critical for describing efficiently coordinated complexity. As we develop a hierarchical model of neural connectionism, we'll discover that each individual hierarchy will map nicely to a meme. And since hierarchies are always built of sub-hierarchies, we'll find that memes are always composed of cooperating sub-memes.

Additionally, we'll discover that this sort of hierarchical organization accurately reflects, much more so than a flat network, the actual physical organization of information flow in the human brain, as measured empirically. From here on, as I talk of connectionist networks in the human brain, I shall be referring to hierarchically organized networks, as I've just described. Now that we understand this hierarchical arrangement, we are prepared for understanding a different style of learning, which we shall next explore.

Hebbian Learning

Neurons don't seem to learn by reward, they learn by the following principle: *Neurons that fire together, wire together.* This style of plasticity, called Hebbian learning, is named after its discoverer, the Canadian neuropsychologist Donald O. Hebb. It is a simple scheme capable of capturing relationships of the world that tend to be reliably repeated – the presence of clouds when it rains, the relative positions of the two eyes in a typical human face, or any other style of pattern that often repeats itself.

Using the mechanism of Hebbian learning, the human brain automatically organizes itself into hierarchical structures, as just described. From the moment a baby is born, it begins to sense many patterns in its environment. When those patterns are transmitted to the brain, they cause coincident firings, which are detected and remembered. The many neurons that fire in response to the detection of coincident relationships among sensory firings, themselves, form inputs to another, higher-level part of the brain that senses and remembers *those* coincident firings. Continuing on in this manner builds structure from the bottom, up, into hierarchies of relationships.

The act of multiple neurons wiring together simply means it takes less signal in such a set of neurons for them to become mutually excited in the future. The sensitization of the neuron set makes them react when they sense the same pattern again, but even more importantly, they also react to an infinite number of *very similar* patterns. We'll consider this process of detecting similarity among patterns in much more detail later.

A child never has to be told how to turn on a light switch, nor be rewarded for it. A child simply notices over and over again that when the switch is flipped, the light goes on. It is a simple relationship that is remembered forever by the child. The human brain is remarkably adept at acquiring and remembering all sorts of relationships, automatically, without even trying to remember, and without any sort of reward.

While the learning of relationships is automatic, the brain does require the concept of reward and punishment to establish some of those relationships as good and some as bad. Our value systems are therefore dependent on how we are rewarded or punished during the learning of various relationships. Realize that all our emotional rewards of pleasure and happiness come, courtesy of our genes, from the recognition of certain patterns that tend to aid in the perpetuation of our genes. And all of our punishments of pain and sadness come also, courtesy of our genes, from the recognition of patterns that have tended to be detrimental to the perpetuation of our genes. Let us be a bit more specific.

The production of pleasure comes directly from the stimulation of neurons that detect such activities as the eating of high-calorie foods, or the having of sex. When those special neurons become active, that neural activity is automatically defined by the brain as pleasure. We'll later discover how many other sources of pleasure can be derived indirectly – learned – from only a few sources of direct pleasure, which are innately defined. At this point, we only need to know that eating ice cream is automatically deemed pleasurable as a result of the brain's ability to detect the intake of valuable energy, which is absolutely necessary for perpetuating genes. Genes have evolved so as to define pleasure as occurring whenever certain conditions related to life-enhancing situations are detected.

We all have innate mechanisms for recognizing certain conditions that have tended to benefit the perpetuation of our ancestors' genes in the evolutionary past. They bring us pleasure and happiness. And we all have innate mechanisms for recognizing certain conditions that have tended to be detrimental to the perpetuation of our genes in the evolutionary past. They bring us pain and sadness. For example, certain neurons are responsible for detecting physical injury, and their firing activity is thus genetically defined as painful. Even when children are scolded by their parents, those children feel displeasure because their genes force them to care about parental approval. Whichever neurons happen to fire from the detection of parental *dis*approval, their activity is defined by the brain as *dis*pleasure. And conversely, whichever neurons happen to detect parental approval, their activity is defined as pleasurable. This is how our genes cause most of us to seek the approval of our parents, especially when we are children, which enables the process of cultural propagation through mimicry.

Allow me to illustrate the distinction between learning relationships automatically and assigning value to relationships on the basis of reward. A young boy, while playing in his front yard, hears two sounds in the distance. One is the sound of a garbage truck working its way through the neighborhood, and the other is the pleasant sound of an ice-cream truck approaching, playing its happy tunes and ringing its bell. The boy knows from experience that there are definite relationships associated with those two different sounds. He has automatically learned both relationships. He knows that soon the garbage truck will pull up to his house and collect his family's garbage. He has learned the particular relationship between the sound and the imminent arrival of the garbage truck without any sort of reward. He simply believes there is no value for him in the arrival of that truck. He also knows that soon the ice-cream truck will arrive at his house. All the times he has been rewarded with ice-cream from that truck cause his mind to assign value to its unique relationship between the sounds that it makes and the eventual taste of ice cream. The point is this: The learning of a relationship is *not* contingent on reward; only the assignment of *value* depends on reward.

Learning is always as automatic as 'wiring together when firing together' – the simple recognition of coincidence. Reward and punishment are only necessary for labeling which of the learned associations are good and which are bad. Think back to Skinner's pigeon. We know it learns when it is rewarded with food. But what if the experiment were changed so that, when the button is pressed, a poker chip is dispensed? The pigeon may still learn the relationship, but would have no motivation to peck at the button again to get another poker chip, because a poker chip has no value to most pigeons (*pigeons* tend to lose their chips quickly in games of poker). We simply could not know, under such a situation, if the pigeon has learned anything, but, given what we now know of Hebbian learning, we must suspect it has.

The difference here is between knowing how to do something and wanting to do it. We humans learn how to do a great many things, but we only ever do what we *want* to do. We do those things that tend to have perpetuation value regarding our defining replicators, for which our genes reward us with happiness. We don't choose the sorts of things that make us happy, our genes do.

We may consider learning as two separate processes, one from the bottom, up, and the other from the top, down. Let us first focus on the bottom-up style of Hebbian learning. Perceptor signals come in from the bottom, and any coincident relationships in those input signals are automatically learned and represented by nodes in the first level of some connection hierarchy. In the very same manner, coincident relationships between firings in those first-level nodes are automatically learned and represented by second tier nodes. The process continues upward to higher and higher levels. Learned relationships build structure from the bottom, up, and it all happens automatically without any sort of reward.

The concept of learning by reward, considered as a top-down sort of learning, could itself actually result from exactly the same Hebbian mechanism of linking simultaneously firing nodes. For instance, suppose some high-level behavioral node is active while the associated behavior is being performed; as an example, consider a node that is active during the behavior of running toward an ice cream truck. And further suppose the behavior does indeed result in pleasure, which eventually causes certain genetically defined 'nodes of pleasure' to be simultaneously firing. An example of a node of pleasure might be a neuron (or set of neurons) that fires during, say, the tasting of ice cream or any other high-calorie food. The simultaneous firings between the high-level node of behavior and the nodes of pleasure will cause them to become linked together. Such linkage can establish a future expectation that the behavior (of running toward the ice cream truck) will result in pleasure.

Some high-level nodes become linked with feelings of pleasure, while others become linked with feelings of displeasure. Those linkages automatically guide our behaviors at any point in time, according to whatever linkage path promises to maximize pleasurable feelings and minimize unpleasant feelings. When we make our decisions, we choose whatever plans are expected to lead toward the excitation of high-level nodes that, from past experience, have become causally linked with pleasure. Just thinking about performing an activity linked with pleasure can cause a pleasurable feeling of excitement in anticipation of its performance.

Even without any reward, we humans learn all sorts of relationships that we store away in memory with the idea that they may be useful someday. For instance, we learn as children that hitting something with a hammer can cause it to break. Then, years later, we might use that knowledge to get the 'meat' out of a walnut or the money out of a 'piggy bank'. We would choose those behaviors only because they are expected to lead toward situations that have become linked with pleasure, such as eating walnuts or spending money on things that are themselves directly linked to pleasure.

While reward is not critical to learning, competition is. We shouldn't have expected any differently if indeed the process of intelligent thinking is, at bottom, evolutionary. We'll ultimately discover at least two levels of competition going on in the human brain. At one level, neural connections compete for strengthening, and, at another level, behaviors and plans compete for getting executed. Yet, we'll find that both paradigms of competition can occur as a result of the simple and automatic mechanism of Hebbian learning.

Behaviors and plans compete for getting executed on the basis of how much pleasure and happiness they are expected to bring. Indeed, we spend our lives pursuing the emotions of pleasure and happiness. So, let us now focus our attention on these aspects of our personalities that so heavily influence our behaviors – our emotions.

Emotions

The primary adaptive purpose of emotions, for both humans and animals, is to switch between various modes of thinking, enabling whatever mode is most conducive to genetic survival in the situation that prevails. So, when I use the term 'emotion', I'll be referring to nothing more than a particular mode of thinking.

Some situations inspire pleasurable emotions, which encourage us to seek out those situations – think of beautiful flowers, lovely fragrances and their natural association with an abundance of food. Other situations inspire unpleasant emotions, which encourage us to avoid them – think of stinking carcasses decaying from bacteria and their natural association with sickness and death. We are automatically encouraged by our emotions to walk toward honeysuckle and away from garbage cans, toward a freshly baked apple pie and away from feces. Indeed, a strong bad smell can often clear people out of a room. Do we volitionally choose to avoid bad odors, or is it fundamentally automatic? I'll try to show that all emotions are (volitionally) uncontrollable; they happen automatically under certain conditions. They form the base motivations for everything we do, and they establish the leveraged mechanisms by which memes are able to impel us into action.

We see a wide range of emotional behaviors in animals. Emotions tend to appear more human-like in the more highly evolved animals, as we would expect from evolution's progress. Fear, hunger and lust are the most primitive emotions, and they exist even in many lower forms of animate life. Higher level emotions, such as loyalty and playfulness, can easily be seen in intermediate forms of life, such as dogs. And the highest levels of emotions, including gratitude and compassion, are seen primarily in the most highly evolved forms of life.

It is important to realize that, for animals, the having of an emotion does not necessarily imply that the emotion causes any type of conscious feeling, as we humans experience. It is terribly reckless to assume that pets consciously feel their emotions simply because they exhibit behaviors that we tend to associate with human emotional feelings. Ray Kurzweil wrote: "It is apparent to me that dogs and cats are conscious (and Searle has said that he acknowledges this as well)" (2005, p.467). While I would expect as much from Searle, I am surprised to read that from Kurzweil. Allow me to illustrate how unfounded that statement is.

What might appear to be emotionally inspired behaviors by pets may simply be adaptive behaviors completely devoid of any associated feelings. As such, purring by cats is no more indicative of conscious feelings of pleasure than yelping by dogs is indicative of conscious feelings of pain. Both behaviors may simply have adaptive value to their genes. I speculate that purring by cats, especially when they are young, simply lets the parents of those cats know when their children are sufficiently fed, comfortably warm, and overall content. Yelping by dogs lets the parents of those dogs know when their offspring are in trouble. Both behaviors – purring and yelping – are inclusively adaptive[7], because they both automatically inspire actions from protective parents that are beneficial to the genes that inspire those reactive behaviors.

The fact that we humans experience conscious feelings of pain on occasions when we make yelping noises does not at all imply that dogs consciously feel pain when they yelp. Perhaps dogs just do what they do, without any feelings of consciousness, because those actions were adaptive in the past. By analogy, we don't suppose a mousetrap needs conscious feelings of intent to kill a mouse. It simply snaps its bar under the right conditions, and, having conscious feelings of intent would not make a mousetrap more effective. Similarly, conscious feelings in animals don't make any behaviors more adaptive, except under certain circumstances that may in fact be unique to humans. I explained at length in *The Laughing Genes* what those circumstances are and why I believe that most animals don't consciously feel their emotions as humans do. A summary goes as follows:

Behaviors are either adaptive or non-adaptive completely independent of whatever conscious feelings might accompany those behaviors. But things change dramatically when there is a cognitive ability for introspectively challenging the reason for executing adaptive behaviors. Once our ancestors developed the ability to ask themselves why they do the things they do, they then needed reasons, associated with benefit, to support their adaptive behaviors. But, if we humans are merely biological machines, then what reason can there possibly be, from the perspective of a machine, for the execution of any behavior whatsoever? A reason implies a benefit, and there is absolutely no way to benefit a machine. Think about it: how can you possibly give benefit to your vacuum cleaner? When our genes made us introspective, they needed to give us hard-wired *reasons* for executing certain behaviors. Those reasons come in the form of pleasures.

We are genetically constructed so as to *believe* we benefit from performing behaviors that actually tend to benefit the perpetuation interests of our genes. So, our reasons for eating, having sex, and so on, manifest themselves as beliefs in feelings of pleasure. They are the necessary links between the interests of our genes and the interests of our individual selves. Metaphorically, the genes deserve a hardy chuckle at how we humans pursue things that we believe benefit ourselves, while the actual benefit accrues to the genes.

Do animals have the ability to cognitively challenge the reasoned basis for their own actions? I suspect that some more-highly-evolved animals might. And, in fact, Kurzweil and Searle may in the end be correct in their assessment of pets; but not necessarily. The important thing to remember is that emotional reactions need not be consciously felt, in order to be adaptive, unless the brain is also capable of introspectively asking itself why it decides to do some things over others. For such an introspective brain, various emotions must be believed to produce either good or bad feelings in order to be motivational. But, whether or not emotions are consciously felt, we only need to recognize that different emotions are designed to put the brain into different modes of operation.

In an obviously dangerous environment, the emotion of fear causes one to seek means of escape, or, at least, to be extremely cautious. There is no question that such an emotion is adaptive if indeed it is inspired by truly dangerous conditions. So the brain must be accurate in its assessment of conditions. This brings up a very important aspect of emotions – how they are triggered. Fear automatically results from the recognition of certain situations. Aggressive animals, angry people with weapons, proximity of fire, and the sensation of falling are but a few of the many situations that inspire fear in humans. All emotions have their respective triggers. They are, in all cases, automatically stimulated by the recognition of certain environmental conditions or situations.

Emotional triggers can involve both external and internal conditions. While fear most often results from characteristics of the external environment, the emotion of hunger results from the recognition of an empty stomach or from low levels of nutrients in the blood, both of which are characteristics of the internal environment. Whereas the emotion of fear puts one in a mode of trying to escape, the emotion of hunger puts one in a mode of trying to find food.

Some emotional triggers are innate, while others are learned. The hunger felt by a baby, and expressed through crying, is surely innate. On the other hand, the fear of flying (more accurately, the fear of crashing) is likely learned through exposure to what the painful consequences can be. But emotional triggers that are learned are no less automatic than those that are innate. Once a valid relationship is learned, it can't be unlearned. Seeing an image of a fiery plane crash changes one's brain forever. The process of learning establishes links between various patterns that are already known by the brain. And those links can extend emotional triggers from innate patterns of sensitivity to learned patterns of sensitivity.

A learned fear of flying – inspired by exposure to a fiery plane crash on television, for instance – links the act of flying to the possibility of painfully burning to death. Such an indirect emotional trigger is just as automatic as the innate emotion of hunger, resulting from an empty stomach. But there is a belief in specific causality – a causal meme – linking the contemplation of flying and the emotion of fear it inspires (for some people). The belief in causality extends the automatic trigger from the fear of fatally burning to the fear of merely flying.

We are beginning to see how the automatic nature of emotions can combine with the automatic nature of learning to give learned memes the automatic means for controlling our brains. My favorite example demonstrating the automatic nature of emotions – one that I presented earlier in *The Laughing Genes* along with much other evidence – involves a man staring at a magazine picture of a naked woman. Here is how Dawkins describes it:

> A man can be aroused, even to erection, by a printed photograph of a woman's body. He is not 'fooled' into thinking that the pattern of printing ink really is a woman. He knows that he is only looking at ink on paper, yet his nervous system responds to it in the same kind of way it might respond to a real woman.
>
> **– Richard Dawkins**, *The Selfish Gene* (1989, p.249)

You can imagine my relief when I read that passage to discover I am not the only man who exhibits this ridiculous behavior. After all, why should I feel lust and a 'growing' measure of physical excitation while knowing full well that I am only staring at a paper magazine? I am apparently not alone in my propensity to get 'horny' at even the *symbolic* presence of a naked woman. The lustful feelings we men get are automatic, courtesy of our genes, so that whenever we find ourselves actually confronted with a real naked woman, we'll be ready and anxious to fulfill our genetic missions.

We automatically feel desires for eating food, staying comfortable and healthy, avoiding pain, having sex, protecting our children, raising families, and earning the respect of others. It is not a coincidence that all those innate desires happen to benefit the perpetuation interests of our genes. Once we accept that all our emotions are automatic and that they establish all our wants and desires, we thereby establish an automatic mechanism for meme propagation. Memes are able to pull the triggers on our automatic emotions so as to cause us to want certain things that we believe will bring pleasure and avoid pain. And we are genetically built to automatically pursue the things we want.

More specifically, memes extend the triggers on our emotions to cause us to want things that are causally related to, and expected to lead to, things for which our emotions were intended. For example, while the male emotion of lust was genetically intended to impel men toward having sex, it can have a much different effect in our modern environment where pornographic magazines are plentiful. Some males buy pornographic magazines because certain causal memes have extended the triggers on their emotion of lust so as to lead those males to believe they can satisfy their lustful desires by buying those magazines. As another example, children get excited at the sound of the ice-cream truck because they have causal expectations – memes of causality – that automatically lead them to believe they can achieve happiness by getting some money and running to the truck.

Indeed, all of our actions are heavily influenced by, if not completely determined by, our expectations of causality. And those critical expectations are learned from experience – experience with pornographic magazines, experience with ice cream trucks, and experience with ... well, absolutely everything.

We are now equipped to understand how culture propagates. Our genes program us with some very special emotions when we're very young, and some other different sorts of emotions when we're parents. Those special emotions ensure that cultural behaviors get passed from parents to children. The emotions to which I am referring are those that give children pleasure when triggered by parental approval and those that give parents pleasure when triggered by observing successful mimicry performed by their children.

Small children usually adore their parents. And they tend to automatically mimic whomever they adore. So, children are naturally inclined toward mimicking their parents. They have emotions that cause them to want to do so. And parents reward their children with shows of approval when the children are successful at doing things that parents do – walk, talk, use a toilet, etc. Parents obviously feel good when their children learn new procedures for healthy living.

There is yet another innate mechanism for ensuring the propagation of culture through parental feelings of sympathy for their children. Indeed, human genes have programmed adult parents so that their babies need not fend for themselves, the parents will fend for them. Human parents are genetically programmed to automatically feel all the emotions that their babies feel, sympathetically, so that the parents will sense the needs of their babies and act to fulfill those needs. Parents live vicariously through their children, sympathetically experiencing their pains and their joys. To know this with certainty, one need only look at the sympathetic concern on a parent's face when its child suffers a serious injury – the parent acts as if it actually feels the pain of its child.

One need only look at the sappy smile on a mother's face as she watches her young child perform in its first band concert, and as she sympathetically feels the pride that the child feels from participating in the production of beautiful music. Of course, the music is not beautiful; it is horrible screeching noise. But the mother acts in sympathy with her child, feeling the child's pride as if it were her own pride and as if the concert were a virtuoso performance, which is how the child perceives it. The mother even feels embarrassment when her child makes an obvious mistake, say, during a solo performance.

By sympathetically feeling the emotions of their children, parents are better able to act toward the satisfaction of their children's survival needs, while at the same time, teaching their children all the cultural norms for reacting to various situations, and for properly interacting with others. A tremendous amount of cultural knowledge is passed to children through observance and mimicry of parents during times when those parents are sympathetically responding to various situations of their children. For example, if you were to ask a very young child how old it is, the *child's mother* is likely to respond "I'm three."

A mother sympathetically responds to various situations encountered by her child, and her child thus learns how to deal with those situations merely by mimicking the mother's vicarious reactions. Both the mother and the child are genetically programmed to automatically do what they do, thereby enabling culture to propagate. They both feel as though they want to do what they do. The child feels joy from parental approval, and the mother feels joy when her child learns how to deal with new situations.

All of our wants and desires result either *directly* from pleasurable feelings, or *indirectly* from expectations of future pleasurable feelings. A direct source of happiness is an activity that produces immediate pleasure, such as having sex or eating tasty food. An indirect source of happiness is an activity or set of activities that is expected to lead to pleasure sometime in the future, such as driving to the store to buy ice cream. An indirect source of happiness is defined by a string of causal memes leading ultimately to a genetically defined direct source of happiness. We tend to believe that by embarking on a plan of going to the store, buying ice cream and then eating it, we will achieve pleasure in the near future. Our minds are built to chase these learned expectations of causality toward future pleasurable emotions. And our emotions of pleasure were evolutionarily designed by our genes on the basis of ancestral situations that have statistically benefited the perpetuation of those genes.

The memetic perspective reveals that a typical human's view of life is really quite 'backward'. A child who races toward the ice cream truck does so, not because he volitionally chooses to, but because he has automatically acquired a meme – a belief – that by doing so he can achieve future pleasure. The sound of the ice cream truck pulls the trigger on his pleasurable emotion of delicious taste, and it does so through his learned expectations of causality.

A man who buys a pornographic magazine does so, not because he volitionally wants to, but because he has automatically acquired a meme – a belief – that by buying the magazine he can achieve future pleasure. The mere thought of the magazine, or the sight of it just sitting there on the rack, automatically pulls the trigger on his emotion of lust, and it does so through his learned expectations of causality. We might say that human males are just pornography's way of getting replicated, as odd as it sounds.

Here is another example of how our typical perspectives are sometimes backward. My family has a little dog – a Pomeranian named Scout. He has an annoying habit of scratching at the door when he wants in or out. And he seems to want in or out every five minutes, or so. There is a spot on both sides of every door where he has worn off the paint. You can imagine my disgust when my mind is deeply engaged in trying to comprehend a complicated book on neural architecture and I suddenly hear that annoying scratching on a door. If I try to ignore it, it will just get more indignant.

Now, we think we keep dogs in order to benefit ourselves, but the truth is quite the opposite. Scout is a lucky animal that is able to survive and thrive with little or no sacrifice. His meals are prepared for him, his home is heated for him, his veterinary bills are paid for him, and his doors are opened for him, whenever he likes. From the viewpoint of evolution, such an animal is a clear winner.

One day, my precocious son said "what if dogs are actually smarter than people, and they have arranged for us to do all their work for them?" It suddenly dawned on me that, while dogs aren't that smart, perhaps their genes are. Is it so wrong to conclude that the genes of dogs have 'discovered' ways to appeal to our human emotions? If you've ever seen a Pomeranian, you'll know just how cute they can be. Perhaps, we humans have 'cuteness' emotions, originally designed by our genes so that we would find our babies to be cute, but now getting triggered by the recognition of many sorts of small dogs.

Indeed, there are many dogs, like Scout, whom I find to be very cute. And there are many other types of dogs I find to be noble in their loyalty. It is certainly the case that we love our dogs because they inspire good emotions within us. And it is not too far-'fetched' to assume that the genes of our domesticated pets have evolved over time so as to appeal to our various emotions. There can be no question that the emotions of humans unwittingly impose strong selection criteria on the genes of domesticated dogs, as we choose to keep some litters and kill others.

We might suspect that the genes of our dogs are metaphorically 'laughing' all the way to the proverbial bank at the way they have evolved so as to manipulate our feelings. And we just keep paying their bills and making them meals. It turns out that memes use us, by preying on our emotions, just as the genes of our dogs use us. All we get out of the deals are good feelings. But on careful reflection, what else is there of true intrinsic value to us, as individuals?

Given that memes are able to prey on our automatic emotions, we now have a complete picture of how and why memes are able to propagate. The ideas in which we tend to have interest are the ideas that excite us emotionally. And those are also the ideas that we are likely to recite to others. We act on ideas that we believe are likely to bring us happiness or avoid pain and discomfort, either now or in the future. We spend our lives pursuing good feelings and avoiding bad ones. Our emotions determine our values, and they impel us into action through our beliefs in causality. Our genes define what we want, and our memes define how we go about getting what we want.

The two critical components to this simple system of motivation are: genetically defined values, and acquired beliefs. But neither our desires nor our beliefs are volitionally chosen by us. They are automatically programmed into us by our genes and our memes of experience. Now that we understand how our genes define our emotions, and hence our values, let us study how we automatically acquire our beliefs.

Beliefs

Grass is green. The sky is blue. It takes a good education to get a good job. Minds are brains and brains are deterministic. Humans evolved from lower species. God doesn't answer prayers. These are but a few of the many things I believe. Intellectually, I am the sum of all my beliefs. Indeed, everyone is intellectually defined by the sums of their respective beliefs. One who believes it is proper to discriminate against black people is a racist. One who believes it is justifiable to have sex with children is a lecherous pedophile. One whose beliefs are all accurate and correct is a very smart person.

My beliefs not only define me intellectually, they also hold tremendous sway over my behaviors. For instance, if I believe that pretty girls lust for musicians, then I'm going to practice on my guitar every day (thinking back to when I was a teenager). If I believe that God answers prayers, then I'll pray every day. But, if I believe that God doesn't like guys who pray for lots of pretty girls, then, instead, I'll pray that I get really good on guitar. I believe he'd like that better.

Everything I decide to do is based on some belief. So, my beliefs guide my actions. But I said the same thing about my emotions. So which is it going to be? Am I guided by my beliefs or my emotions? The truth is, they work in conjunction to guide my behaviors. I do whatever I *believe* will bring me the most pleasant *emotions*. My genes define my automatic emotions, and my memes define my beliefs.

The reason that boys tend to like pretty girls is because they believe they'll get a lot of pleasure from being with them. Whether or not their expectations turn out to be valid, it is what they believe that guides their behaviors. The reason we all like to eat tasty food is because we believe it to be a pleasurable experience. But we also hold other relevant beliefs about tasty foods. For example, we believe that fattening foods will bring us misery in the long run, from being fat or from generally poor health. So, there is often a trade-off between short-term and long-term expectations of pleasure.

We often speak of *willpower* as the ability to forego short-term pleasure. But what we don't often consciously realize is that we are only willing to sacrifice in the short term if it is in exchange for the expectation of much greater long-term pleasure. Since expectations can only arise out of beliefs, our beliefs are instrumental in the enabling of willpower. After all, we need to hold the respective expectations or beliefs that our sacrifices will indeed be rewarded in the future.

Notice that we acquire our beliefs automatically. Even though we feel as though we volitionally choose some of our beliefs, such cannot be the case. Firstly, experiential evidence *is what it is*, and my senses would not let me believe grass is any other color than green. I am forced to believe whatever my senses tell me, completely independent of what I might 'choose' to believe. Indeed, all my beliefs must be consistent with the things I sense. Secondly, my beliefs regarding things I can't sense, such as my beliefs regarding evolution, have come mostly from people I trust. I tend to trust scientists more than I trust preachers. And when I was young, I tended to trust my parents above everyone else; both of them were firm believers in evolution.

Consider an interesting belief that I now happen to hold on the basis of neither sensation nor trust. I believe that all minds are deterministic. And I strongly hold that conviction despite opposing beliefs by nearly all of my friends, family and associates. Do I *choose* to believe that? Yes, I do, but my choice is automatically determined by the operation of my brain circuits, over which I have absolutely no volitional control. My heredity in conjunction with my total history of experiences have forced me to 'choose' that belief. My brain has automatically determined such a belief to be the only one consistent with all my other beliefs that have automatically been acquired on the basis of sensation and trust.

The human brain automatically assesses the validity of any new belief, to which it is exposed, by checking its consistency with old beliefs. It is a completely automatic process – not a matter of choice. While trying to think of a supporting example, it occurred to me that if you question my assertion, then you have inadvertently proved my point (especially if you don't remember volitionally deciding to do so). On the other hand, if you believe my assertion, then I need say no more. Human brains have evolved so as to automatically assess the validity of beliefs, simply because beliefs are so critical to the mechanical process of intelligence.

Beliefs are indistinguishable from causal memes. Even though many of our beliefs are acquired from experiential evidence, rather than by mimicry, beliefs are memetic patterns nonetheless. And they serve as templates for their own replication or translation in several ways. Firstly, we may think about a belief over and over again. And that act of re-considering a belief essentially replicates it. Secondly, we are able to communicate a belief to others, and that act also replicates it, via multiple translations. We might initially have trouble finding a specific pattern associated with any given belief, but there will surely be a pattern of neural firings in my brain that gets replicated every time I think about a certain belief. And there will be some specific patterns of words that get replicated every time I communicate a given belief.

We'll later see quite clearly that a belief is a pattern of relationships. It does not really exist as a specific and fixed pattern in either space or time. A belief merely has to maintain its overall structure of hierarchically organized relationships between many real-world spatio-temporal patterns, as it gets itself replicated. The same belief will take on different specific neural connection patterns in different brains, but in all cases it will maintain the same relationships between real and conceptual objects. Conceptual objects are built from hierarchical relationships among real objects, and all real objects are nothing more than sensed patterns.

A belief is like a multi-tiered 'mobile' hanging above a baby's crib. The shape of the mobile can change with the breeze, but the relationships between hanging objects and the arms of the mobile remain invariant. For example, the common belief that grass is green takes on a completely different pattern of neural connections in different brains, but in all cases, it relates the concept of grass with the concept of green. And the concept of grass can be further reduced, as it relates blade-like shapes with attachment to the ground. The color green is indicated by specific photoreceptors that are sensitive to light of a particular frequency range, and the blade-like shapes are indicated by various spatial patterns of firing photoreceptors being simultaneously excited in the retina of the eye.

At the bottom of all relationship structures are real-world sensations of patterns. So, our beliefs are hierarchical relationships built up from concurrently sensed real-world patterns. Even the beliefs that we acquire through communication by language are characterized by high-level relationships between already-existing hierarchies of relationships that map all the way down to real-world patterns. In essence, a sentence merely relates various words, and those words represent patterns and relationships with which we are already familiar.

Behavior Based on Causal Expectations
Our human brains certainly have significant vestiges of our animal heritage that compel us toward immediate gratification. But our brains also have uniquely human capabilities of predictive planning, which, when coupled with a willingness to defer gratification, allow us to invest in the future. That ability to plan ahead is enabled by the automatic collection in memory of all sorts of causal relationships. Long-term planning intelligently produces brand new, never-been-tried sorts of complex behaviors, simply by stringing together familiar, simple behaviors.

Whereas vestigial parts of the human brain were designed to operate on the basis of conditioned behavior (as exhibited by Pavlov's dog), the neocortex seems to have been designed to search for whatever will be expected to bring reward in the future. Whereas conditioned behavior chooses on the basis of whatever led to reward in the past, the act of planning intelligently asks: What sorts of causal expectations can be strung together to get from the current state to a future state that is likely to involve reward? The conflict between these two strategies is manifest as the dilemma of willpower. The human mind struggles between its vestigial urges toward immediate gratification and its expectations of greater future reward through immediate sacrifice.

To do planning, we don't need the reinforcement effects of reward; what we need are a multitude of various sorts of nuggets of causality that we might use in some combination to change the environment from its current state to some more pleasurable future goal state. Whether we realize it or not, our brains constantly analyze the world by looking for, and remembering, valid nuggets of causality – causal relationships at many levels of abstraction. Those nuggets of causality are memes. Whereas our genes program us to seek paths to happiness, those paths are always constructed from causal memes that we cognitively string together.

Recall the previous description of memes as having input and output states. For a given meme, the relationship between its input and output states is defined by the nature of causality captured within the meme. If the output state of one causal meme matches the input state of another causal meme, then the two memes may be connected into a two-step plan. As a frivolous example, consider the following situation: If my emotional state is currently dominated by hunger and my goal is to eat some food, then I must search my library of causal memes for any that allows me to do that. What I find is a meme representing the act of reaching into an open refrigerator and grabbing some food. But what if the refrigerator is closed? Well, another search of my causal meme library yields one that represents the act of pulling on the handle of a closed refrigerator to open it. Now I have a two-step plan built from two causal memes having matching interfaces. The output state of one meme produces an open refrigerator, which is the required input state of the other meme.

The hard part of stringing together causal memes into plans is the identification of memes having complementary interfaces. Like doing a jigsaw puzzle, the hard part is finding the pieces that fit together. Snapping them into place is easy. At some point, we need to fully understand what it takes for two memes to be complementary. For instance, in the two step process just described, what was it about those causal memes – opening the refrigerator, and reaching inside to grab food – that allowed them to be coherently joined together?

We will eventually discuss precisely what it is that defines a measure of complementarity between two memes. For now, let us simply think of memes as having various attributes that define the ways they are able to fit with other memes, sort of like the funny shapes on the sides of jigsaw puzzle pieces. But the characteristics of memetic interfaces are far more complex than those of just shape. Memetic interfaces are additionally characterized by a great many abstract attributes, including color, sound, smell, texture, hardness, when, where and for what purpose. We'll later discover that memetic interfaces are abstractly represented by nested relationships of many sorts of patterns existing in multiple dimensions.

Through a very simplified visual model, we may think of cause-and-effect memes (causal memes) as being something like dominos, each having a pattern of dots on one end and a different pattern of dots on the other end. We are able to lay the dominos end to end so long as adjoining ends have matching dot patterns. To implement a plan, we must find a chain of dominos that allows us to get from the pattern that represents what currently exists to the pattern that represents the goal of the plan.

Now, keeping that simple model of causal memes in mind, recall the discussion regarding direct and indirect emotions. Direct emotions are defined by genes; they yield their associated feelings immediately during the performance of such activities as having sex or eating tasty foods. We may suppose that those activities are genetically defined as pleasurable, simply because they have statistically tended to perpetuate the genes that define them that way. Indirect emotions, on the other hand, result from having a bunch of causal memes of a specific type. Those causal memes 'connect' to our emotional triggers and thereby extend those few triggers, which are genetically defined, to a great many more that are indirectly defined on the basis of various causal relationships we have learned. These indirect emotions automatically create our wants and desires, thereby leading us down paths that are likely to propagate our genes.

Indeed, we experience many sources of pleasure and happiness that don't fall neatly into the fundamental categories defined by our genes. But in all cases, they can be seen as steppingstones on paths expected to lead to the satisfaction of genetically defined fundamental sources of happiness. For instance, a young man in college might find himself very happy when he gets the grade of 'A+' on a test. But his happiness results indirectly from his subconscious expectation that getting good grades in college will eventually allow him to get a good job, which will lead to lots of money, which will enable him to eat well, buy nice clothes, and attract a fit girlfriend with whom he might have sex and eventually start a family. Every stepping stone on the path toward the perpetuation of his genes represents an indirect source of happiness for him.

All our short-term goals, such as trying to do well on upcoming tests, are completely defined by causal expectations that link them as if they are steppingstones on some path to real future happiness. We get excitement at the prospect of achieving those short-term goals, and a dose of 'indirect' happiness from actually achieving them. Our feelings of excitement and anticipation automatically form our intentions, and impel us to act.

In this manner of making us excited over the anticipation of future happiness, our genes lead us down paths that tend to benefit their perpetuation interests. We dutifully follow those paths with rarely any conscious understanding of why we do so. All we typically know is that we will be rewarded with pleasure or happiness as we successfully reach various stepping stones along the paths. No wonder the genes have good reason to laugh, metaphorically. And, no wonder the memes have good reason to humorously mock us, metaphorically, for our robotic behaviors – behaviors that are in fact inspired by them, the memes.

While animals tend to live 'in the moment', by impulsively following their immediate emotions, we humans tend to live in the future, chasing the emotions that we anticipate from cause-and-effect relationships. Genes relinquished a great deal of control to memes, simply because memes carried so much survival value back in the days when hominids traveled in families, and memes were thereby propagated mostly vertically. But now, memes are able to use the anticipation mechanism as a means for propagating themselves, either vertically or horizontally, simply by citing a string of cause-and-effect relationships that make us believe we will be rewarded in the future.

Religious memes, for example, are able to tickle a trigger of excitement by making some of us imagine the pleasantries of heaven. Religions cite the relationship of working hard now and playing by the rules of the religion in order to get into heaven after death. Such a motivation would not be possible without a willingness to defer gratification into the future, even beyond death. We must be ever vigilant against non-valid causal memes that play us for fools by promising future payoffs that don't actually exist, such as those payoffs that are promised to occur after death.

As another example, liberals and conservatives are both concerned with maximizing their future benefit, but they choose different causal paths to justify their expectations. Liberals advocate helping the down-trodden, now, so that they themselves – the liberals – will be helped if ever in a similar down-trodden position. Conservatives rationally advocate saving as much as possible, now, so that they themselves will never be in a position of needing help. In both cases, the immediate sacrifice of helping or saving is expected to yield potential payoff in the future.

By the way, liberals need not be consciously aware that the underlying basis for their compassion is derived from expectations of future reciprocity or a similar sort of benefit. They may indeed feel emotionally concerned with only the well-being of the unfortunate objects of their compassion. But evolution is not in the business of breeding emotions that are devoid of adaptive benefit. Even though liberals may feel emotions that are purely compassionate, they only have those emotions because their ancestors tended to benefit from them in some way.

Humans are compassionate only because compassion tends to yield some sort of benefit – a return on investment – in the long run. That benefit tends to result from the fact that most humans also feel emotions of gratitude, which they direct toward those from whom they have received acts of compassion. So, a compassionate act is indeed an intelligent investment whenever it is directed toward one who is likely to feel gratitude and is therefore likely to reciprocate the help when it is needed in the future.

The principle of investment – sacrificing in the short-term for greater future benefit – is a characteristic of all intelligent behavior, including even some behaviors produced by biological evolution. Indeed, many animals engage in practices of investment, none more obvious than spiders that invest in the construction of webs. Most animals embark on investment strategies as a result of genetic programming. I doubt that spiders know why they bother to weave their webs; they just do so because they are genetically programmed to do so. Indeed, all animals that expend effort in the construction of things like webs, dams, nests and hives, and even in the process of hunting or gathering food, are unwittingly investing in their futures. They do so as a result of their genetic programming – on the basis of things learned by their genes before birth.

Humans, on the other hand, make investments on the basis of intellectual modeling of the future, using causal memes learned by their brains after birth. The tremendously important principle of investing toward the expectation of greater future benefit forms the basis for our modern business capital markets and our employment compensation practices: Work today, get paid on Friday. The principle of investment has even been codified by our societal memes in the form of contract law. As I previously mentioned, all forms of cooperation require up-front sacrifice – investment – by someone. And, since cooperation is the essence of life, it should not be a surprise that our cultures have evolved so as to facilitate it. Our brains certainly have.

The human brain is genetically designed to automatically calculate a measure of expected happiness for every option available to it. The option expected to yield the most future happiness is the one that is automatically chosen. But, while this mechanism may seem simple on the surface of it, there is a complication arising from the various timings associated with when various receipts of happiness are expected to occur. For example, if options A and B both yield identical happiness in exchange for identical effort or sacrifice, then the choice is a toss-up, unless their payoffs of happiness occur at different times. For identical receipts of happiness occurring at different times, the earlier is always preferable.

The situation becomes crystal clear when we consider money as a proxy for happiness. We may logically do so because money can buy many things that lead directly or indirectly to happiness. Now, consider a situation in which I have an opportunity to sell my car to either buyer A, who offers $1000 immediately, or buyer B, who also offers $1000, but won't deliver the cash for a full year. In both cases, I sacrifice my car immediately, and in both cases I gain $1000. But buyer B's intention to pay me a year later forces me to accept some added risk in the deal. I might die before I receive his payment, or he might go bankrupt in the meantime. Most people will intuitively prefer buyer A over buyer B.

Financial analysts have learned how to quantify the risk associated with future cash flows in a theory they call the *time-value of money*. Future cash flows are discounted by a degree correlated with how far in the future they are expected to occur. In essence, near-term cash flows are weighted more heavily than long-term cash flows. Recognize that the differences in weightings can vary between investors. For instance, a ninety-year-old investor may place a higher value on a quick payoff than would a much younger investor. The various weightings, averaged across all investors, determine the aggregate interest rate for a particular waiting period on otherwise 'riskless' U.S. Treasury bonds. Returning to our example, the interest rate for a particular individual can be determined by asking: How much more would buyer B have to pay (a year later) so that his offer would be preferred?

When a human brain makes a decision between two options, involving expected payouts of future happiness at different times, it must assign different weights to each option depending on their timings. We may call this concept the *time-value of pleasure*. Just as different investors have different preferences regarding the timings of returns, so do different people have different preferences regarding the timings of expected happiness receipts. Some people are willing to work hard, now, for the promise of greater future rewards, while others just want to be immediately happy.

The moral dilemma represents the trade-off between a willingness to invest effort toward greater future happiness and the more juvenile desire for immediate gratification. Moral behavior always requires immediate sacrifice toward the promise of greater future reward, while immoral behavior gives way to immediate gratification. Doing something nice for a neighbor, which entails an immediate sacrifice, earns a 'chit' of reciprocity that can later be cashed in during times of trouble. It is the moral thing to do because it facilitates cooperation in a way that requires immediate sacrifice, yet it is only rational (and moral) if it is expected to yield future benefit that exceeds the immediate sacrifice.

We tend to think of moral behavior as selfless behavior, and in the near term it is. But in the long term, moral behavior tends to be even more selfishly justified than immediately gratifying behavior. Yet, the selfish justification of moral behavior need not be realized at the conscious level. That is, a nice guy need not realize that the justification for his niceness results only from expectations of grateful reciprocity. So long as the beneficiaries of invested sacrifices – the recipients of moral behaviors – are likely to experience feelings of gratitude, then the sacrifices of moral behaviors can yield future benefit, and thus be adaptive.

Notice that we don't typically direct our generosities toward nasty people who have demonstrated a reluctance to ever reciprocate acts of kindness. We tend to be kind toward others who are also kind, and likely to be grateful. In that manner, our generous investments in kindness are rarely wasted – they are indeed likely to yield future return.

Our genes have programmed us to perform unsolicited acts of kindness for others simply because our genes evolved in environments where most people were similarly programmed to feel gratitude in response to acts of generosity toward them. Those feelings of gratitude usually led to reciprocation of kindness and generosity when it was most needed. The emotions of compassion and gratitude got the principle of cooperation started, but intelligence is what fully empowered it. Intelligence has evolved as a means for exploiting the possibilities of achieving greater future benefit in return for the investment of near-term sacrifice. And cooperative relationships always require sacrifice by someone.

Emotions were first engineered by genes to inspire *reactive* behaviors, by putting the brain into a mode appropriate for whatever environmental conditions happen to be sensed. Intelligence was then engineered by genes to enable *proactive* behaviors. By enabling the anticipation of future pleasurable emotions through causal expectations, we humans have become imbued with excitement and intentional purpose. But recognize that intent can never be volitional. It is always automatic and involuntary. We modern humans routinely go off on purposeful missions, sometimes involving great initial sacrifice or investment of effort, but always promising to yield even greater benefit in the form of long-term happiness. While we individuals believe we benefit from being happy, the true benefit of happiness-producing situations tends to accrue to the perpetuation interests of our genes.

Before any intelligent human ever decides to make the sacrifice of an investment, it had better have a reasonably certain expectation of a greater future return of benefit. Such a reasonably certain expectation amounts to a prediction of the future. Indeed, the essence of intelligence is the ability to predict the future under various circumstances. But how are we able to predict the future? What is it about the structure of our human brains that allows them to collect and use memes of causality so as to act as clairvoyants? We are slowly and methodically closing in on the answers to those questions. We will get there eventually. But the next chapter will diverge slightly, into a philosophical discussion of how it can be the case that many diverse components – such as collections of neurons – are able to achieve the sort of synergy that leads to the production of a coherent mind.

Chapter 6. **The Synergy of Mind**

Before we set out to contemplate the architectural characteristics of the human brain, which will be the next chapter's dedicated pursuit, we must first develop a confidence that there can exist some sort of a synergy that enables an intelligent mind to emerge from a plethora of much simpler neurons. After all, how is it possible that mechanistic neurons can be coordinated so as to represent our many beliefs, and how can our many disparate beliefs interact so as to form our intelligent and conscious minds? And, what, exactly, constitutes a conscious feeling? This chapter focuses on trying to gain some insights into answering those provocative questions.

We have already begun to conceptually decompose much of intelligence into a few tractable processes. One is the Hebbian learning process that automatically acquires beliefs about the world. Another process automatically connects complementary beliefs of causality together into various coherent plans, from which the single best plan is identified as the one yielding the most expected benefit of future pleasure or happiness. While we have made significant progress toward discovering the essence of intelligence, we still need to determine, at some philosophical level, what it means to *understand* a concept. All those issues will be fully explored in later chapters.

This chapter will prepare us for upcoming discussions by examining the synergy that can emerge from a group of human minds. The emergence of a synergistic sort of collective wisdom from a group of minds is important and relevant as an analogous means for understanding how a single human mind can synergistically emerge from hundreds of billions of mechanistic neurons.

Beyond Collective Wisdom

It is now widely accepted among cognitive scientists that what we humans feel as a unitary consciousness actually results from many specialized modules. Collectively, those modules achieve the synergistic production of a single intelligent mind. When I speak of modules in this context, I am not referring to such large-scale components of the brain as the amygdala or the hippocampus or the thalamus. I use the term 'module' here in reference to an elemental unit of cortex roughly corresponding to the neural representation of a concept, or meme. We'll later see that such a module is composed of a somewhat scattered but connected group of neurons in the cortex. The size and structure of such a neural ensemble will become evident in the next chapter.

As a very loose analogy of how intelligence can be modular, consider a kitchen that is equipped with a mousetrap, a strip of sticky fly-paper, an electronic bug zapper, a roach motel (roaches check in but they don't check out), an ant trap, and a cache of moth balls. We are entitled to say that the kitchen is very 'intelligent' at preventing pests. Such a style of intelligence results from a simple accumulation of knowledge. Each module 'knows' how to prevent a particular type of pest. Together they 'know' how to prevent many types of pests.

Strictly speaking, the term 'synergy' implies a product that is greater than the sum of its constituent parts. And while our first example of modular intelligence only accumulates knowledge, later examples will show that the level of intelligence can go up geometrically with respect to the level of knowledge, especially when there is cooperation involved among the modules. Indeed, a simple accumulation of knowledge can sometimes produce a synergistic rise in intelligence – a level of intelligence that didn't previously exist. The synergy relates to the huge number of cooperative combinations that are possible from the knowledge of only a few types of components. For instance, with only the knowledge of a few types of basic electronic components (transistors, resistors, capacitors and inductors) we are able to produce millions of useful electronic products, simply by combining them in different ways. Now, let us see how synergy relates to modularity.

Another example of many modules cooperating toward the production of intelligence is evident in the way carpenter ants are able to find the sticky honey jar in my kitchen. They work together through a simple mechanism of coordination. Random exploration by many ants guarantees that one of them will eventually discover the jackpot. That lucky ant simply lays down a pheromone trail when he leaves, which guides other ants to the valuable source of food. Considered together, they act as coordinated modules of a bigger brain. The coordination among them synergistically enhances the likelihood they will find a good source of food. And it guarantees that most of them will benefit from the inevitable discovery. Just as a group of ants can synergistically act more intelligently than a single ant can act, so might we suspect that a huge group of neurons can act far more intelligently than a single neuron can act.

The idea that a mind can synergistically emerge out of inferior components strikes at the heart of John Searle's famous philosophical argument, commonly referred to as the *Chinese Room*. Searle asks us to imagine a small room having a window through which questions, written in Chinese, can be handed to an English-speaking man inside. The man, having no understanding of Chinese, has all the proper instructional material for looking up the Chinese characters in big books of tables and for producing a different string of Chinese characters to be written down and handed back out through the window. Searle asks us to assume the responses are syntactically and semantically appropriate to the same extent that a typical Chinese person's answers would be. To a Chinese person outside the room, who submits questions and receives answers, the room appears to 'understand' Chinese. Yet, amazingly, there is nothing in the room that understands Chinese.

Searle believes this simple thought experiment shows that there is no way an automated system could ever experience a true understanding of Chinese questions, or anything else, for that matter. According to Searle, computers will never understand the information they automatically process. But, I suggest that Searle's argument shows only that 'understanding' does not occur at the component level. It occurs, instead, as a synergistic product of the aggregate system. I will try to show that a conscious feeling of understanding can indeed emerge from components that have no such feelings. This theory will become especially credible when we eventually reduce conscious feelings to their essence. We'll return to Searle's Chinese Room during a later discussion of 'understanding'.

One of my goals is to prove that a collection of neurons can know and understand and even *feel* things that are beyond the knowledge and understanding and feeling of each individual neuron. It is the synergy of the many neurons, acting cooperatively, that allows complex knowledge, understanding, and even consciousness to emerge. More generally, we need to fully appreciate how cognitive synergy can indeed emerge from many cooperating components, even though the components themselves have no means for the sort of cognition that is achieved by the aggregate system.

We have to be careful how we use the concept of synergy. For example, I will eventually show exactly how conscious feelings can emerge from collections of neurons, but we may not likewise infer that perhaps *free will* can similarly emerge from many neurons, each of which is governed entirely by natural laws. There is no sort of logical argument to support the idea that collections of interacting deterministic devices, such as neurons, might possibly transcend causation through some sort of synergy. Synergistic effects are always reliably repeatable, and therefore, completely deterministic.

So that we may better understand how the synergy of many interacting neurons can cause a human brain to become an intelligent human mind, let us consider a similar sort of synergistic effect occurring at a higher level – at the level of many brains working together. It turns out that the synergy emerging from many people working together allows them, as a coordinated group, to greatly exceed their individual capabilities.

Here is the underlying premise: In large groups of communicating people, language memes enable the emergence of a meta-mind. Very simply, language allows a society to convey to all its members a set of coordinating conventions – its rules and laws. And, such a set of conventions can enforce a style of cooperation that enables synergy to emerge. For instance, the coordinating conventions of ants are very simple, yet they enable a group of ants to continuously evolve batter paths to sources of food. The language of communication between ants is built on pheromone trails rather than words. Ants use their pheromone trails as symbols for paths to food, just as humans use various words as symbols for all sorts of things. Bees similarly coordinate their efforts by communicating paths to food. Bees use various body movements as symbols for various directions.

By establishing conventions for cooperation, the collective mind of a society achieves a sort of synergy that, amazingly, allows it in some cases to 'know' things that none of its members knows. Additionally, these conventions allow the collective mind of a society to discover solutions to problems more quickly. Just as a large group of ants can evolve better and better paths to food, so can a large group of humans evolve better and better answers to difficult questions that are way beyond the capabilities of most individuals. Let's see how this can happen.

It is easy to see how the collective knowledge of a group, coming from the combined knowledge of the group's members, can exceed the knowledge of any particular individual in the group. For example, no single person at NASA knows how to build a space shuttle, but the collective group of all the employees at NASA (and its subcontractors), taken in aggregate, knows how. While there certainly can be an accumulation of knowledge among the members of a cooperating group, which does allow the group to be more knowledgable in aggregate, there is much more to synergistic intelligence than that. There is a sort of synergy that can emerge, when conditions are right, that allows a group to actually think better than even the sum of the thinking over all of its individuals (assuming duplicated thoughts don't contribute to the sum).

Intellectual synergy arises from the way information is exchanged among group members and is finally aggregated. Humans appear to have social habits that were evolutionarily designed to exploit a group-level style of cognitive synergy. A group of humans effectively explores problems by having various members propose solutions that are then evaluated by everyone in the group. Solutions are thus able to evolve by this group-level mechanism of iterated trial and evaluation. We'll eventually discover that the neurons of a human mind are indeed organized so as to facilitate a very similar evolution of thought, completely within a single brain.

When a group of people contemplates some decision, the product of the collective contemplation can be vastly different from what any individual in the group would produce on its own. This curious phenomenon allows us to speak of a group as having a mind of its own. We'll see that the collective mind of a group can indeed be much better or much worse than the minds of its individual members, but first, we need to define what it means to have a 'better' mind. Given a problem having a definite answer, a better mind is one that statistically produces a decision that is closer to the correct answer. So, whether a group mind is better or worse than the individual minds of its members can only be assessed with regard to problems having definite answers. There are such problems that we can use to see whether a collective mind is better than its individual minds. We'll find that the performance of a group mind depends on how the group is organized. That is, it depends on what rules and conventions the group uses for aggregating the minds of its individuals.

Perhaps the most common method for a group to organize its collective thoughts is through the principle of compromise. We'll soon see that compromise usually yields a better collective mind for decisions involving a single variable. But when confronted with a decision involving multiple variables, compromise can be a disaster. Take, for example, a government managing the two variables of taxation and spending. The U. S. government, being mostly composed of Republicans and Democrats, must aggregate two vastly different styles of individual mind. Republicans claim to want low taxes and low spending, and Democrats claim to want high taxes and high spending. There are logical arguments in support of each strategy as to how they provide long-term benefit to society at large. But the way in which Republicans and Democrats often choose to compromise makes no sense at all.

Here's the deal: Republicans agree to give-in to Democrats on the issue of spending, and Democrats agree to give-in to Republicans on the issue of taxation. There ... problem solved ... compromise to the rescue. No, problem not solved. What we are left with is high spending and low taxation. Even though everyone is happy now, the long-term consequences of the compromise are likely to be detrimental to the unified society. Notice that, through compromise on issues involving multiple variables, the group can decide on a strategy that is vastly different from any of the strategies preferred by the individual members of the group. A group really can have a collective mind of its own.

We must take care when extolling the virtues of compromise to limit its use to single-variable decisions only. Such a limitation makes compromise a useless strategy for many real-world decisions, but nevertheless, there are some simple decisions for which compromise is absolutely appropriate. For simple problems involving a single variable, it turns out that compromise does indeed tend to yield a smarter group mind than most or all of the individual minds that comprise it. We may contemplate a thought experiment that illustrates just how this can be the case.

Consider the trivial task of estimating the number of jellybeans in a sealed glass jar. The task is trivial in the sense that the outcome doesn't matter very much. But the concept is applicable to important situations as well. Given a sealed glass jar full of jellybeans, there are several scientific ways of estimating the number inside. While you may be thinking of a method yourself, it is unlikely to be the one I'll propose, which is very simple in theory, yet largely ignored.

The most common scientific technique might be to estimate the volume of a single jellybean and divide that volume into the measured volume of the jar. There are several problems with this method, mostly involving the way jellybeans pack themselves together. Nevertheless, one could probably estimate to within 20 or 30 percent of the actual number.

Now, let me propose a method for estimating the number of jellybeans in the jar that is theoretically more accurate and amazingly simple, albeit a bit unwieldy. The surprising veracity of the simple technique became clear to me while reading James Surowiecki's book, *The Wisdom of Crowds* (2004). All we need to do is show our jar of beans to a diverse crowd of independent people and have each of them apply their best technique for guessing the number. A simple statistical analysis shows that the average of all the guesses can be made arbitrarily close to the actual number by simply using a larger crowd. Here is how the theory goes:

If we were to choose a fellow randomly off the street and ask him to take his best guess at how many beans are in the jar, we can mathematically model the prediction of his response as a bell-shaped probability distribution centered around the actual number of beans in the jar. This is a fortunate situation, because, if it is precisely true that our mathematical model of an individual's response is accurate, then we can easily find the actual number of jellybeans in the jar simply by finding the mean of the distribution.

Now, I am forced to admit that there may be some built-in bias that would cause the distribution to be centered around some number other than the actual number of beans in the jar. For instance, it may be the case that humans systematically under-estimate the sight of volume. After all, a doubling in volume can be achieved by increasing each of the three dimensions by only about 25%. But I am not preventing anyone from actually measuring the jar. In any event, while a bias may in fact exist, the theory I'm about to describe still serves to show how greater intelligence can emerge from components of lesser intelligence. And, I'll later discuss how systematic bias can be eliminated, or at least minimized. So, let us proceed, for the moment, under the assumption that there is no such bias that exists across the average of all participants.

We return to the task of determining the mean of the supposed probability distribution, so that we can, in theory, know the precise number of jellybeans in the jar. But how can we find the mean of the distribution? Well, since the same bell-shaped probability distribution applies to all people, we can find its approximate mean by asking a few people and averaging their responses. So then, we can get a better approximation of the number of beans in the jar – better than we could get from a single person – by asking several people to estimate, and then averaging their guesses. The more samples we take – the more people we ask – the more accurate our approximation will likely be. Now, here is the magical part: Theoretically, we can come arbitrarily close to the actual number of beans by asking more and more diverse and independent people to take their best guess. By asking a thousand diverse people to guess the number of beans in the jar, and averaging their guesses, we can expect to come very close to the actual number, in theory.

I find this phenomenon somewhat surprising. It clearly shows that a large group of people can collectively know the answer, even though no single person in the group knows the answer. This is a clear example of the effect that John Searle thought to be so unexplainable. If a crowd of people can 'know' things that are unknown to all the individuals in the crowd, then perhaps, similarly, a room can 'understand' Chinese even though there is nothing in the room that 'understands' Chinese.

This simple phenomenon of collective wisdom was first noticed by the statistician Francis Galton in 1906 when he witnessed a weight-judging contest at a county fair. There were 787 participants who submitted their guesses as to how much a fat ox on display would weigh after being slaughtered and dressed. Galton asked for the submission slips after the contest was over. He was surprised to find that the average of all the guesses missed the weight of the ox, which was actually 1198 pounds, by only one pound. The collective wisdom of the crowd guessed the weight of the ox almost perfectly, much more accurately than many experts did.

We shouldn't have expected any differently. If there is any sort of probabilistic connection at all between the actual weight of the ox and a person's guess, and if we may assume there is no systematic bias, then each individual guess can indeed be described by a probability distribution having its mean at the actual weight. After all, it is logical to assume that as a reasonable fellow makes his guess it is equally likely that he will guess too high as too low. So, as many reasonable people make their guesses, their deviations will cancel out, and the average of all those guesses will theoretically converge toward the *exact* number. When Galton realized the statistical implications of his discovery, he quickly became aware of a more significant implication in the domain of social governance.

The act of many people voting in a democracy is a mechanism of compromise by which a group can discover its average guess as to who is the better candidate. And, when there are only two candidates in an election, there is really only one variable to consider. The variable captures how much better or worse candidate A is compared to candidate B. Every vote can be expressed mathematically as +1 if the vote is for candidate A, or -1 if the vote is for candidate B. The average consensus is calculated by counting up all the individual contributions. If the sum is greater than zero, then candidate A wins; if the sum is less than zero, candidate B wins. Elections can be expected to yield good results, so long as there is only one variable to consider. But when there are more than two candidates in an election, there are multiple variables to consider, and the compromise arrived at through voting is less likely to yield a good result.

Can a diverse group of voters, most of whom are uneducated, be trusted to successfully pick their own leader? Wouldn't a society be better served if an elite group of highly educated individuals were responsible for making all the decisions regarding societal conduct, including the selection of their successors? Perhaps an elite group should be appointed to govern the society in perpetuity. In fact, when Galton collected the contest submissions, he had expected to find support for the idea that experts could guess the weight of the ox much more accurately than the group as a whole. Galton held a belief that the masses were not up to the task of collectively managing their own fates. But after analyzing the contest submission slips, he was surprised to find that the *group* as a whole did better than the *individual* experts in the group. How could this be the case? Shouldn't we expect *individual* experts to do much better than *individual* non-experts? Yes, but amazingly, it does not logically follow that a *group* of experts will collectively do better than a *group* of non-experts. Let us try to understand why.

Recall that the only theoretical caveat preventing a large group from collectively determining exactly how many jellybeans are in a given jar relates to the possibility of a systematic bias across all the group members. It turns out that a group of experts is more likely to have such a bias than a diverse group of non-experts. I am indeed surprised to discover that the performance of a group's collective decision-making may be improved by excluding experts in favor of a more diverse group of non-experts.

The Value of Diversity and Independence

When assembling a group to work on some problem, it may be more important to have diversity in the group than to load the group with expertise in the problem. This concept, often discussed by University of Michigan professor Scott Page (in reference to an upcoming book titled *The Logic of Diversity*), is strongly counterintuitive to most people. Indeed, many smart people often wrongly believe that experts always know better than the rest of us, even collectively. The name for such an improper opinion is *elitism.* It turns out that the rest of us, when taken in aggregate, are simply less likely to suffer from the individual biases that tend to plague experts.

We clearly see the value of diversity over elitism in the aggregate stock market, in which the prices of stocks are determined by a very wide array of opinions from many very unsophisticated investors. We might expect that an expert investor should do much better at picking stocks than the average investor. But in fact this doesn't appear to be the case. Studies consistently show that professional fund managers typically under-perform the overall market.

Now, an under-performing fund manager might absolve himself on the basis that the current values of his portfolio stocks are determined by the same pool of unsophisticated investors that mis-valued them when he bought them. And so, an expert stock analyst might claim that his under-performance results from the self-fulfilling effect of the overall group of investors continuing to mis-value his stocks. But the same phenomenon, regarding an expert's inability to perform better than the average, can be seen in other domains where outcomes are ultimately determined more objectively. Horse racing is a prime example. The performance of each horse is initially speculated on by the gamblers, but is ultimately determined by the outcome of the race. There can be no self-fulfilling effect in horse racing.

The value of diversity is clearly revealed by the typical outcomes of racetrack betting. It is very difficult, even for a professional racetrack gambler, to beat the odds at the track, even though the pool of bettors is diversely distributed over many types of unsophisticated bettors. Some are professionals, but most are there just for amusement. The toteboard odds reflect the average opinions over all bettors, mostly non-experts. And yet, they tend to predict almost perfectly how likely it is that a horse will win. According to Surowiecki, statistical evidence shows that 3:1 horses win a third of the time, 4:1 horses win a quarter of the time, and so on. Favorites win most often; second-favorites win second-most often, etc. Statistically, the aggregate crowd 'knows' the finishing order, all the way down the line, even before the race is run. The crowd at the racetrack predicts the future more accurately than the experts do, at least probabilistically. The fact that professional gamblers can't reliably beat the racetrack odds is strong evidence that the accuracy of those odds comes from the diversity of the bettors.

So, why should it be the case that crowds are smarter than experts? Well, in my experience with stock analysts, I find that they all tend to share similar opinions about the stock market, at any given point in time. It should not be a surprise that their opinions become synchronized, because their individual opinions tend to originate from opinions that they have received from other stock analysts. The discussing of opinions between experts forms a self-reinforcing source for systematic bias among them. Whatever bias exists at the outset gets exaggerated by this effect. I believe such bias does naturally emerge among of all sorts of collaborating 'experts', including stock analysts and professional racetrack gamblers.

There is yet another source of bias among experts. Returning to the jellybean example, if there were such things as experts in counting jellybeans – let's call them 'bean counters' – they very well may carry a systematic bias related to the fact that experts are only deemed 'experts' when they all learn the very same accepted techniques for performing their art. These sources of bias among experts invalidate the assumption that the bell-shaped curve of likelihood regarding their guesses is centered at the actual number of beans in the jar.

Experts think deterministically just like non-experts do. But the thinking of experts is deterministically biased in a way that non-expert thinking is not. So, here then is another precaution we must take to ensure proper group thinking. We must avoid the possibility that the thinking of individuals in the group is in any way synchronized or inter-dependent. The best way to do that is to ensure diversity.

I know of a particularly interesting example of wrong expert bias that I am happy to share. The example comes from an issue of the magazine *Skeptical Inquirer* (vol. 15, Summer 1991), which talks of a brain teaser submitted to a reputedly high-IQ columnist, Marilyn vos Savant, in the weekly magazine *Parade*. It goes as follows:

> Suppose you're on a game show, and you're given a choice of three doors. Behind one door is a car; behind the others, goats. You pick a door – say, No. 1 – and the host, who knows what's behind the doors, opens another door – say, No. 3 – which has a goat. He then says to you, "Do you want to pick door No. 2?" Is it to your advantage to switch your choice?

In her column, Marilyn vos Savant responded affirmatively, and explained that the original odds of 1-in-3 go up to 2-in-3 by switching to door No. 2. She is exactly correct, and she clearly explained how and why she is correct in succeeding articles. Yet, she received the following chastising letters from 'experts' who should know better:

"I'm very concerned with the general public's lack of mathematical skills. Please help by confessing your error. ..."

– Robert Sachs, Ph.D., George Mason University

"You blew it, and you blew it big! ... You seem to have difficulty grasping the basic principle at work here. ... There is enough mathematical illiteracy in this country, and we don't need the world's highest IQ propagating more. Shame!"

– Scott Smith, Ph.D., University of Florida

"Your answer to the question is in error. But if it is any consolation, many of my colleagues have also been stumped by this problem."

– Barry Pasternack, Ph.D., California Faculty Association

"I am in shock that after being corrected by at least three mathematicians, you still do not see your mistake."

– Kent Ford, Dickinson State University

"...Albert Einstein earned a dearer place in the hearts of the people after he admitted his errors."

– Frank Rose, Ph.D., University of Michigan

"...Your answer is clearly at odds with the truth."

– James Rauff, Ph.D., Millikin University

"May I suggest that you obtain and refer to a standard textbook on probability. ..."

– Charles Reid, Ph.D., University of Florida

"...I am sure you will receive many letters from high school and college students. Perhaps you should keep a few addresses for help with future columns."

– W. Robert Smith, Ph.D., Georgia State University

"You are utterly incorrect. ... How many irate mathematicians are needed to get you to change your mind?"

– E. Ray Bobo, Ph.D., Georgetown University

"If all those Ph.D.s were wrong, the country would be in serious trouble."

– Everett Harman, Ph.D., U.S. Army Research Institute

"Maybe women look at math problems differently than men."

– Don Edwards, Sunriver, Oregon

If you find yourself agreeing with the arrogant 'experts', allow me to enlighten you with the explanation Marilyn vos Savant gives:

> Suppose there are a million doors, and you pick door No. 1. Then the host, who knows what's behind the doors and will always avoid the one with the prize, opens them all except door No. 777,777. You'd switch to that door pretty fast, wouldn't you?

Michael Shermer gives another good explanation for the same problem in his book *The Science of Good and Evil* (2004, p.170), which I'll paraphrase: There are three possible situations for what exists behind the doors: (1) car, goat, goat; (2) goat, car, goat; or, (3) goat, goat, car. Assuming you choose door number one, and the host subsequently reveals a goat, then, in possibility one, you lose by switching, but in possibilities two and three, you win by switching. You'll find the same is true no matter what door you choose.

Here is the way I prefer to explain it: If your strategy, going into the game, is to always switch doors whenever a goat is revealed by the host, you will win whenever you initially choose a door with a goat behind it (you choose a goat, the host reveals the other goat, and you switch to the car). Thus, your probability of winning will be 66%. But, if your strategy is to always stick with the originally chosen door, then you will win whenever you initially choose the door with the car behind it, thus, your probability of winning will be 33%.

Now, the critical question is: Would a group of experts be more or less likely than a diverse group of non-experts to get this problem right? I envision a group of intelligent mathematicians coming quickly to a decision on this problem based on what appears to be its obviousness, whereas, a diverse group of intelligent people who are not mathematicians might indeed agonize over it for a long time. After all, the act of finding something to be obvious requires a confidence in one's own expertise on the matter. At the very least, a group of experts had better be diverse enough to include some who are willing to challenge their own intuitions.

While the value of diversity is clear, the benefits of diversity can be nullified if, for whatever reason, the diverse members of a group tend to synchronize. Unfortunately, such synchronized thinking often prevails in groups, and yields a '*herd mentality*' that usually produces very bad decisions. Weaker members can often be coerced, or at least heavily influenced, by stronger members. And so, groups of decision makers tend to operate more effectively if their opinions can be gathered *before* they are influenced by social pressures to conform.

We have established that groups of people think most effectively when they are diverse and independent. We will later see that the same principle applies at the neural level. Neurons automatically divvy themselves up into modules, representing concepts, so that they are as diverse and as independent as possible. The divvying-up into 'orthogonal' concepts occurs by way of a nonlinear neural mechanism involving lateral inhibition.

The value of independent diversity is also quite obvious in various instances of evolution. Indeed, diversity of species is critical to the progress of biological evolution. For example, the Irish potato famine was caused by over-reliance on a single strain of potato that succumbed to a single strain of blight. At the other extreme, even though dinosaurs went extinct from some catastrophic event, the diversity of life at that time allowed mammals to take their place. So, the idea that diversity of opinion is critical to creative thinking and proper decision-making is in strong agreement with the theory that human intelligence reduces to an evolutionary process.

Certainly, many of the great ideas in history were considered ridiculous or outlandish before they became widely accepted. Great ideas never come from things that are obvious to everyone; they always come from things that are obvious to only one or a few individuals. It takes a frothy, chaotic mix of wild ideas among a group to fully explore all the possible alternatives. It takes diversity to ensure that the unique individual with the great idea is included in the group. And, it takes proper rules of coordination among the members of a group to ensure that the great idea will be encouraged to emerge.

Having established that groups of diverse and independent people are able to achieve cognitive synergy through proper coordination of their thinking, we now shift the focus of our memetic perspective from the inter-brain phenomenon involving the evolution of culture, and such synergistic effects as the emergent wisdom of crowds, down to the intra-brain evolution of mental activity and the emergent wisdom of neurons. We must suspect, at this point, that intelligent minds are those that are full of diverse ideas and beliefs. In fact, we'll discover evidence suggesting that human consciousness actually seems to emerge from such a large and diverse collection of beliefs.

Conglomeration of Beliefs

Is a human brain already wired for consciousness at birth, or could it be the case that consciousness is built almost entirely out of things that are learned? The idea that consciousness is learned, first proposed by Daniel Dennett, seemed to me, for a very long time, odd and unlikely. But I am now a believer. I'll argue in support of the astonishing hypothesis that all our beliefs (most of which are learned), taken in aggregate, are responsible for all our human intellectual capabilities, including even our experiences of consciousness.

With the exception of a few beliefs regarding our own pleasurable and painful feelings under certain conditions, we humans aren't born with any beliefs at all; we learn them. As we are exposed to a wider and wider range of sensations, and as we learn more and more about how the world works, we thereby build up many beliefs. That enormous set of beliefs is sometimes referred to as a memeplex. I'll try to show that our learned beliefs enable us to become aware of ourselves and how our selves relate to the rest of the world. Such is the nature of consciousness. Gradually, over time, by learning all sorts of recurring patterns, and relationships among those patterns, our sensations become perceptions.

Certainly, a very small set of conscious determinants are innate. Indeed, the few innate beliefs with which a baby is born are obvious to parents. A baby believes it feels hunger under certain conditions, discomfort or pain under certain other conditions, and pleasure when being fed. Those few innate beliefs in feelings of pleasure and pain persist throughout one's entire life (recall that our genes give us those beliefs, and a few others, so that our actions will be consistent with their interests). And a few more innate conscious determinants, in the form of beliefs in feelings, are genetically programmed to hormonally develop later in life. They are related to sex, social ordering, and the various feelings of parenthood. But, a great many more conscious determinants will be learned through life experiences.

Let us contemplate the learned determinants of consciousness by considering a thought experiment similar to one I previously described.

143

Suppose we could take a new-born human baby and hook it up to an intravenous feeding machine that would nourish the baby for, say, twenty years. Further suppose we deprive the subject of all visual and auditory sensory stimulation by keeping it in a dark and silent room. What would the twenty-year-old subject experience, consciously, when it is finally released from its 'cocoon'?

We might be tempted to believe it would experience the same sort of consciousness that any normally-raised twenty-year-old would experience. But such cannot be the case, because it takes repeated exposure to common patterns before Hebbian learning can develop the means to distinguish between any of them. Without any history of exposure to coherent sensory patterns, a human brain will remain a completely unorganized, haywire mishmash of randomly connected neurons. The conscious experience of such a brain might be something like looking through scratched-up glasses, of the wrong prescription, on an extremely foggy day. Light will get through, but any sense of pattern will be lost.

Babies are not born with the ability to discriminate between shapes, not even between a square and a circle. Presented with two squares and a circle we must suspect that a baby would be unable to even discern which two patterns are alike. The ability to discern between visual patterns is acquired through Hebbian learning applied to firing neurons of vision as those neurons are exposed to various recurring elements of patterns, such as edges, corners and arcs. Perhaps I can illustrate the learned ability to discriminate among patterns through analogy.

Suppose you close your eyes and touch some Braille writing. Would you be able to discern between the characters? Many blind people can, but only after repeated exposure. Now, I am not referring to the ability to know which set of dots corresponds to which character; I am more fundamentally referring to the ability to merely distinguish between various sets of dots. You may label them in whatever way you want.

Suppose I play many pieces of classical music for you that you have never heard before, all of which were composed by either Bach or Mozart. Would you be able to group them by composer? A good musician could, but only after significant exposure to many pieces by each of the two composers. Suppose I give you many glasses of red wine, all of which are either a Shiraz or a Merlot. Could you properly group them? A wine connoisseur certainly could, thanks to many experiences with each type of wine. Could you tell the difference between Folgers and Maxwell House coffee? If so, would you have been able to at the age of five? Some people can tell the difference, but only after having experienced many samples of each. Which do you prefer: Coke or Pepsi? A milk drinker would never know the difference, but to someone who drinks a lot of different sodas, the distinction is easy to make.

We *learn* to discriminate between various patterns of shapes, patterns of sounds and patterns of tastes. We *learn* language and the conventional relationships between words and various patterns of shapes, sounds and tastes. For example, wine connoisseurs use phrases such as: "The wine has a dry, nutty, fruity, oaky sort of flavor." Such a combination of flavors is only discernible after significant exposure to each flavor individually. We *learn* concepts as relationships of words. And we get better at understanding those concepts with repeated exposure to real-world instances of them.

Here is the salient point: We experience life by classifying everything in terms of patterns already learned. To one who has never been exposed to any patterns, one has no basis on which to classify new patterns, and therefore, one's experience of life is extremely nebulous. Patterns enter the senses and are quickly diffused into random neural firings that are full of wrath and fury, but signifying absolutely nothing.

It appears that the mind is a conglomeration of intertwined, yet highly organized patterns. And those learned patterns capture all the mind's beliefs. So, the mind is a conglomeration of learned beliefs. It will later become perfectly clear, when we discuss neural architecture, that all our experiences of memory amount to patterns of neural firings that tend to repeat the firing patterns initially experienced when the memories were formed. Indeed, all learned firing patterns serve as templates for their own replication or translation – the very definition of memes. Our learned experiences are replicated whenever they are recalled by memory. And they are replicated, via multiple translation, whenever they are conveyed to others through language. All our learned experiences are memes.

Now we can appreciate Dennett's description of consciousness as "a huge complex of memes (or more exactly, meme effects in brains) ..." (1991, p.210). Susan Blackmore takes the idea a little bit further:

> Memetics provides a new way of looking at the self. The self is a vast memeplex – perhaps the most insidious and pervasive memeplex of all. I shall call it the 'selfplex'. The selfplex permeates all our experience and all our thinking so that we are unable to see it clearly for what it is – a bunch of memes. ... By acquiring the status

of a personal belief a meme gets a big advantage. Ideas that can get inside a self – that is, become 'my' ideas, or 'my' opinions, are winners. ... In conclusion, the selfplex is successful not because it is true or good or beautiful; nor because it helps our genes; nor because it makes us happy. It is successful because the memes that get inside it persuade us (those poor overstretched physical systems) to work for their propagation. What a clever trick. That is, I suggest, why we all live our lives as a lie, and sometimes a desperately unhappy and confused lie. The memes have made us do it – because a 'self' aids their replication.

– **Susan Blackmore**, *The Meme Machine* (1999, pp.231-234)

She is implying the existence of some cultural memes, common to most of us, which make us believe we are much more than just deterministic biological robots. Our self images, as freely acting agents, are fabricated by these memes. They coerce us into adopting various compatible memes, such as religious or political or moral beliefs, as major constituents of our personalities. They then give us beliefs in the importance of proselytizing others into believing them. They even stand to perpetuate more readily by making us believe we should fight for what we believe is right. But why would a biological robot do such a thing, unless it believes it is something more than just a robot?

It turns out that a transcendent notion of 'self' is a critical element for some aspects of our memetic conscious experiences. For instance, how could anyone revel in the emotion of love, while simultaneously believing that one and one's lover are both just robotic protection vehicles for their respective genes? But, by embracing the idea of a transcendent self, we also open the door to the belief in life after death, which can lead to such radical behaviors as suicidal terrorism.

Now, think back to that twenty-year-old raised in a dark and silent cocoon. It should now be clear to you what a blank slate it would have for a brain. Not only would it have no words to describe a square or a circle, it would have no ability of pattern recognition to even discriminate any difference between them. It would have no sense of self, and everything would look and sound similarly vague.

It seems that, as babies, we humans have only an extremely limited sort of consciousness, but as we mature into adults, our sense of awareness grows in proportion to how much we have sensed and how much we know.

Kids are clueless, but on the other hand, they experience life in terms of concepts that are much starker in contrast. Issues are simple for them – black and white – only because they haven't yet learned all the subtle nuances. And sometimes their simple viewpoints are more accurate than our cloudy, muddled, confused, adult perspectives. We can't help thinking in indiscernible shades of gray, given the multitude of partially conflicting, barely accurate memes we are fed throughout our lives. And it is a rare individual adult indeed whose bed of knowledge contains nothing but accurate memes, thereby allowing tall towers of starkly-defined abstract thoughts to be built on top of their solid foundation.

Consciousness

It is a quaint characteristic of most humans that they think they possess the ability to experience something a machine could never have – a conscious experience. But if the human mind is just a very complex machine, then a machine must be capable of having a human-like conscious experience. While the logic of that statement is simple and straightforward, we simply cannot imagine how machines, such as computers, could possibly feel consciousness. Let us now try to reconcile our intuitions with our logic.

The experience of human-like consciousness requires one more ingredient, beyond a hotbed of memetic beliefs. This final ingredient is likely to emerge from a mind's natural tendency to reconcile all its many beliefs. Consciousness requires, after the memetic development of a 'self', an ability for *self-reflection*. This entails the ability to question one's own reasons for doing the things one does. Dennett describes the typical human self-image as a "benign user illusion." He claims that the image of one's self as a conscious entity results as a natural consequence of a machine monitoring and internally representing its own state, including, not just its physical state, but also its mental state. Allow me to clarify this idea of *self-monitoring* with an analogy, using a corporate organization to represent a mind.

Suppose a politician were to place a call to the editor of the New York Times complaining about yesterday's story, which focused on corruption by that very same politician. The editor, having a copy of yesterday's newspaper on his desk, reads the article. By so doing, the organization of the New York Times is effectively monitoring its past actions. If the editor then inquires as to whether the author of the article is in the building, the organization is effectively monitoring a part of its own current state.

This ability for self-reference apparently forms the basis for consciousness. Such is the claim of philosopher David Rosenthal (1986, 1989, 1990a, b). Dennett (1991, p.210) summarizes Rosenthal's view this way: "... when a mental state is accompanied by a conscious or unconscious higher-order thought to the effect that one has it, this *ipso facto* guarantees that the mental state is a conscious state!"

We know that the human brain has transverse connections running all over the place. The enormous amount of inter-connectivity within the brain amounts to significant self-reference, thereby allowing self-awareness, and introspection. The entire brain can be well-aware of events for which only a part of it is responsible. When one part of the brain makes a decision to act, another part can analyze that decision in a context different from the one in which the decision was made. In fact, the decision can be considered from many different perspectives.

I am now convinced that much of the conscious experience results from learning. Consider our feelings of free will. Perhaps they come from our parents telling us such things as: "If you're good then Santa Claus will give you what you want and you'll eventually go to heaven, but if you're bad then Santa will put coal in your stocking and you'll eventually go to hell." Every time they say such things, they reinforce our beliefs in our own abilities to freely choose whether to be good or bad.

Human minds are programmed to think in ways that statistically perpetuate the genes that design them. This is an undeniable implication of the evolutionary process. And another undeniable implication logically follows, as I previously mentioned: Even if free will were possible in a mostly deterministic universe, such a style of thinking would be less desirable from evolution's perspective than a style of thinking that deterministically seeks to maximize the chances of perpetuating the genes that enable that thinking. Any talk of a volitional style of free will is completely incompatible with what we know of evolution. We might as well talk of spirits.

There are perfectly natural reasons for our wanting to believe in spirits. One of those reasons relates to how we model other people. Since we can't know their life histories, the states of their brains, nor their genetic make-ups, we must model them as unpredictable agents. That supposed unpredictability is indistinguishable from free will. Since we see ourselves as similar to other people, then, for consistency, we must likewise model ourselves as freely acting, unpredictable agents. There are indeed many other reasons for our wanting to suppose spiritual forces. Let us consider some of them.

Have you ever stood next to a backyard fire, or a campfire? Why is it that, no matter on what side of the fire I stand, the smoke always seems to blow toward me? Is there some spiritual force at work in the smoke? It reminds me of when I'm at the grocery store and, no matter in which checkout line I stand, it always seems to be the slowest one. What is the force that so obviously conspires against me?

Of course, there is no spiritual force; the pattern I perceive is automatically conjured completely within my own mind. Any belief of mine that some sort of force is conspiring against me is just an inaccurate belief. Here is an example of how our incorrect beliefs come to exist: Whenever I stand next to a fire and the smoke is blowing away from me, I think about other things. The only time I ever think about the direction in which the smoke is blowing is when it is blowing directly in my face. So, when I think back to all the times I've ever considered the direction of smoke from a fire, every one of those times it was in my face. No wonder I come to believe the smoke has a mind of its own.

Similarly, when I stand in a very slow checkout line, it tends to capture my attention. I focus my thoughts on how slow it is. On the other hand, when I am in a briskly moving line, I tend to think about other things. So, when I think back to all the times I've ever considered the speed of my checkout line, everyone of those times my line was slower than the other lines. No wonder I come to believe there is a higher mental power conspiring against me. No wonder I often hear people saying things like: "*with my luck*, ... I'll probably be struck by lightning," as if their luck is worse than average. Just as some people believe they have bad luck, or that the forces of nature are conspiring against them, so do we all tend to gather similar inaccurate beliefs about our feelings of consciousness.

Whenever we reflect on our own thoughts, there is no doubt that we believe we feel consciousness. When we think back to all the times we've ever thought about our own thoughts, every one of those times we merely believed we were conscious. So, it is not surprising that we come to believe our thoughts represent some sort of inner spiritual essence of what we call consciousness. Indeed, Susan Blackmore once wrote an article speculating on "why consciousness only exists when you look for it."

Our human minds are essentially systems of many beliefs. Our cognitive processes do nothing but manipulate beliefs. Beliefs are the currency of thought. Our minds automatically string together causal beliefs into plans, or expectations, that end in feelings of pleasure and happiness. But, what can account for those conscious feelings of pleasure? How can a biological machine have such things as feelings? The answer is quite simple: It can't. All it can have are beliefs. And yet, I am quite certain that, at various times, I *feel* such things as pleasure and pain, hot and cold, and the tastes of sweet and sour. At least, I *believe* I have those feelings of sensation.

If beliefs truly are the currency of thought, then perhaps it is simply more accurate to say that we *believe* we feel consciousness than to say we *actually* feel consciousness. Indeed, it seems evident to me that a mere belief in a conscious feeling is all it takes to make the feeling real. Think about it. Is there any possible situation under which you can believe you feel pleasure without actually feeling it? No, 'feelings' and 'beliefs in feelings' are fundamentally identical.

This explanation takes on credibility when we acknowledge the temporal aspects of our feelings. Consider, for example, our occasional thoughts about how we felt at some point of time in the past. Most people would agree that such a reflection on memory necessarily involves a belief. For instance, I believe I remember being sad yesterday ... or it could have been the day before. As it turns out, we can only believe in the things we remember, including even the feelings we remember having. Yet, the same must be true when we think about how we are feeling right now. There really is no such thing as *right now*. The present is defined as a singularity – a point that consumes no span of time whatsoever. What we think of as the present moment is actually a span of time that exists entirely in the past – the recent past, but the past nonetheless. So, when we think about how we are feeling in the present moment, such a thought is really a reflection on the recent past, which can only be represented by a belief.

Our genes have discovered that our beliefs in our own feelings are all we need in order to link the goals of ourselves, as individuals, to the goals of our genes. They, our genes, reward us with pleasure and happiness when we do things that tend to benefit their chances of perpetuation, and they punish us with pain and displeasure when we do things that tend to threaten their chances of perpetuation. Our conscious feelings, then, amount to nothing more than innate beliefs that become active under certain conditions. For example, in a man's brain there is an innate belief in a feeling of extreme pleasure that becomes true when his genitals are stimulated during the act of sex. That same belief in extreme pleasure can even be made true under the artificial circumstances of masturbation, so I've been told. The brain simply activates its own belief in a feeling of pleasure whenever it recognizes the specific neural activity that results from genital stimulation.

If, during sex, a man's sex partner asks him if he is enjoying himself, his brain will respond affirmatively. And, if the man, by thinking about his own thoughts, asks himself whether he is feeling pleasure, his brain responds affirmatively. Regarding every aspect of his cognitive processes, he completely believes he is feeling pleasure. As it turns out, his belief in the feeling makes it real. Indeed, there is absolutely no way for him to believe he is feeling pleasure without actually feeling it.

Most of us aren't typically aware that our genes provide us with innate desires for sex, food, comfort, healthy children and social esteem. We don't really appreciate what motivates us. When we think of why we do the things we do, we simply *feel* as though we want to do those things. But where do those 'wants' come from? Do we volitionally conjure them out of thin air? That's what we tend to believe, even though it is clearly not true. So, even our wants and desires can be expressed as beliefs. We simply believe we want certain things. Just as feelings and desires can be reduced to mere beliefs, so can causal memes also be reduced to beliefs. Let's see how:

The mental process of stringing together elements of causality – the process of planning – amounts to stringing together a bunch of beliefs. This becomes perfectly obvious when we realize that all our plans are based on our beliefs. For example, a kid goes to college because he *believes* he'll be able to get a better job as a result. And he further *believes* that by having a better job he will be able to achieve more future happiness. Those causal beliefs form his general plan for life.

Most of the time, we go through life responding like robots to events that happen around us. We react according to how our genes and memes have programmed us to react: They force us to automatically identify and choose whatever courses that we believe will tend to maximize total long-term happiness. Sometimes the proper courses aren't clear, and we automatically simulate various different options for a while – we mull them over. By mentally evaluating many such plans, and remembering the best, we have a computational means for maximizing future reward. It is indeed an evolutionary process, operating within a system of beliefs.

It seems that all of cognition is concerned with manipulating various types of beliefs: beliefs in feelings, and beliefs in causality. It seems that beliefs are indeed the currency of thought. And, as it turns out, beliefs are easy to implement computationally. In fact, a computer database is nothing more than a bunch of beliefs. We'll soon see how the brain mechanistically encodes its beliefs.

No matter how sure you are that you actually feel pleasure, as opposed to simply *believing* you feel pleasure, your confidence in that feeling is still just a belief, although a very strong belief. This is a bitter 'pill' for many people to swallow, but recognize that a bitter pill only tastes bitter if one who takes it believes it tastes bitter. And further recognize that a good hypnotist might be able to convince you that a bitter pill tastes as sweet as candy simply by manipulating your beliefs. In similar fashion, completely as a result of my heredity and environment, I have become convinced that the 'pill' - the robotic nature of the human mind – is not bitter at all. Allow me to elaborate:

Many people apparently believe that being a robot is a bad thing, because that is what they have been taught to believe. Because of that belief, they experience displeasure at the suggestion that they themselves are biological robots. That displeasure prevents them from accepting the obvious truth about themselves. I, on the other hand, happen to believe that being a robot is a good thing, because I am convinced that only a deterministic mind can have the opportunity for exhibiting rational thought and reliable behavior. The more deterministic a mind is, the more rational and reliable it can be. My odd belief makes me perfectly content with who and what I am. My many unique beliefs allow me to fully reconcile my self-image with everything I know about science and nature. Believe it or not, I get a comfortable feeling of higher awareness from my scientifically consistent beliefs.

Chapter 7. **The Mimicking Mind**

We have finally reached the heart of the matter. Having learned all about mimicry and how it is destined to emerge in the evolutionary ascent of all life, everywhere in the universe, we are now at a point where we need to contemplate an architectural framework for describing how the process of mimicry maps onto the neurons of the human cortex. By what sort of neural processes does a brain record the behaviors of others and then re-apply those behaviors to its own body at appropriate times?

This is risky business. Given the enormity of the task, and the many degrees of freedom involved, whatever I propose will be much more likely to be at least slightly wrong than absolutely right. But we have to start somewhere.

Our task of understanding human cognition is made significantly easier by the recognition that all cognitive elements, at their essence, are expressible as beliefs. All our motivations are genetically provided in the form of beliefs in our own conscious feelings, and all our methods for pursuing what we believe to be pleasurable feelings result from our beliefs in causality. We'll find other similar sorts of simplifications that allow us to unify various cognitive effects on the basis of commonality at their essence.

Causal Memes

The ultimate goal of this book is to determine how a human brain comes to 'understand' the world in which it exists. How can it drive a car, play a guitar or know what exists at the center of a star? The situation gets a whole lot more intriguing upon the realization that a brain cannot see, hear, smell, touch or sense the world in any way other than by a flurry of electrical impulses – spikes – arriving via bundles of neural fibers. The brain knows nothing of odor, or melody, or shape, or texture, or color. Those characteristics are all transduced – converted to spikes – at the level of the senses, well before they reach the brain. Spikes in ... spikes out. That's it. That is all the brain knows. This fact is amazing in itself, because it suggests that a brain operates only on the basis of firing rates or patterns that it recognizes in its incoming neural activity.

Given what we know about the operational nature of a neuron, its gross functional characteristics have to be pretty simple. With respect to its many inputs, it seems that a neuron is only capable of integrating firings from them or detecting coincident firings across multiple instances of them. The model I'll propose for cortical neurons relies much more heavily on the process of coincidence detection than on integration. Such is the means for facilitating the previously discussed mechanism of Hebbian learning.

The recognition of coincident firings is, apparently, the primary function of each individual element in the cortex. Coincident firings between two neural fibers can signal a real-world relationship of some sort. The most valuable relationships to recognize and learn are ones that link spatial patterns across time. They represent causality. We should suspect that brains are designed specifically to learn such causal relationships.

I have referred to *causal memes* many times already as recurring patterns of behavior having rather abstract yet well-defined input and output states. Being memes, they must serve as templates for their own replication or translation, but it has not been at all clear so far how they might do that. I must now bring some rigor to the definition of causal memes in order to describe how they are stored and used by the brain. So, let us try to organize our thoughts on this subject.

Allow me to define here, just for the sake of organizational convenience, a new class of meme existing in some sense above the behavioral memes that are responsible for culture. As an indication of their higher status, we might refer to these new replicators as *hypermemes*. Whereas cultural memes are behaviors that serve as templates for replication by others, hypermemes only *represent* those behaviors in cognitive processes. For instance, when I perform a ritualistic behavior, it is a cultural meme, but my contemplation of that behavior is a cognitive hypermeme. My ability to contemplate that behavior over and over again is an indication of that hypermeme's ability to serve as a template for its own replication. And, my ability to convey that hypermeme to others through language is also an example of it serving as a template for its own replication, via multiple translation.

What I am referring to as a hypermeme is somewhat similar to the static version of what Robert Aunger speculates on in his book *The Electric Meme* (2002). But, whereas Aunger only defines his electric meme in the vaguest of terms, we'll be much more specific in the way we define the hypermeme. We'll see exactly where it comes from, what its boundaries are, how it is encoded in the neural architecture, how it mutates, and how two hypermemes are complementary when they share significant neural structures in the representations of their connecting interfaces. Of course, we'll need to discuss some details of neural organization and functionality. But, thankfully, the brain achieves its complexity primarily through redundant expressions of relatively simple structures.

Toward supporting the claim that hypermemes are the sole currency of thought, we might be tempted to think of them as representing real-world objects and real-world behaviors, because we are certainly capable of contemplating various objects as well as various behaviors. But, by extracting the essence of what it means to experience something – either an object or a behavior – we find that we may generalize to a single sort of underlying representation. Let us think of hypermemes as representing only dynamic behaviors, with the recognition that one style of behavior is that of remaining still. So, stationary objects are still dynamically encoded, but in terms of their mundane behaviors of acting motionless.

Our brains are more like movie cameras than like still cameras. And, when I say the word 'tree', you may initially conjure up a very still tree on a calm, windless day. But if I also tell you a storm is brewing and the wind is howling, your mental image of that tree will change dramatically. Your mental image of that tree was always capable of moving, even though your initial impression of it may have been motionless. Objects can only be useful to us by virtue of the way they interact with other things. And interaction always involves a temporal component of motion. So our mental representations of things are like movies; and, for each and every object, we hold many movie snippets under various conditions, of which there will likely be one that is completely motionless.

So, by modeling all objects, including static ones, as if they are always capable of being dynamic, we may contemplate using them in various ways so as to direct energy toward the perpetuation of our defining replicators. We may conjure up a still image of an apple tree, but what is really important is how apples may be easily plucked from it. We may conjure up a still image of an apple, but what is really important is how the apple interacts with a mouth that bites into it. Can you almost taste the apple as you imagine biting into it? If so, then you must have a previously acquired movie snippet that encodes all aspects of the dynamic nature of biting into an apple, including even the emotional reaction of pleasurable taste. Objects are never important. Only our dynamic activities of interacting with objects are important.

Our genes have designed our brains so that mental patterns of causality can be strung together to form plans that will lead to enhanced perpetuation for our genes. By picking an apple, and then biting into it, I may enhance the probability of my genes being perpetuated. It is a simple, two-step plan built from hypermemes of causality that I've automatically collected in the past. We may now be completely clear as to what causal memes are. We usually talk of them as behaviors, but those behaviors are always translated to and from patterns of neural activity – hypermemes. As we use the term 'causal meme' from here on out, let us keep in mind that it fundamentally refers to behavioral patterns that have been translated to neural firing patterns. Now that we understand what causal memes actually are, I'll abandon the use of the term 'hypermeme'.

The question that now confronts us is how we might architecturally describe these causal memes in a way that is consistent with what we now know of how the brain is wired, and in a way that allows us to automatically identify and connect their complementary interfaces. As it turns out, the brilliant inventor Jeff Hawkins has recently published a wonderful book called *On Intelligence* (2004) that proposes a neural connection model having all the architectural elements we need. He refers to his model as the *memory-prediction* framework.

The Memory-Prediction Model of Mind

We are finally ready to describe the functional architecture of the human brain – a brain that constantly gathers statistics about how the world works, and automatically organizes its many billions of neurons into connecting patterns that reflect the natural laws, properties and repetitive relationships of the real world. Those connecting patterns form a dynamic model that learns to simulate the characteristics of local reality.

Given that the brain is intended to model the real-world for the purpose of predictive planning, let us return to the idea that the brain can only know what it knows by way of electrochemical impulses arriving via bundles of nerve fibers. From those flurries of neural spikes the brain must construct a mental representation of the sensed external environment, faithfully representing, in some manner, all sorts of real-world relationships and object properties. Hawkins suggests a theory for doing just that. And, lucky for us, the theory accomplishes its magic by effectively transforming the brain's architectural complexity into architectural redundancy. In the typical style of nature's elegance, there appears to be a universal algorithm in the brain for processing all sensory data. After careful reflection, we should have expected as much, because there just aren't a lot of metrics to be algorithmically compiled over a single fiber's train of spikes, or even over all the spikes carried by a bundle of fibers. Allow me to clarify.

Judging from evidence regarding the operational characteristics of a typical neuron, it is highly unlikely that a single neuron encodes rich serial data in the temporal pattern of its firings. Neurons do not send a 'Morse code' style of message, which requires the clocking of a synchronous signal, as a modem does. A neuron quite simply appears to convey a single measure of signal strength through the frequency of its firings. And so, we may exclude sophisticated temporal encoding schemes, such as the synchronous signals found in modern-day electronic communication devices, from the possible mechanisms by which neurons communicate. But then, what is the essence of communication among neurons?

Think of it this way: Suppose you and I are separated by a great distance, and there are three strings running between us, call them A, B and C, that I may use to send you messages. If I pull on string A, it conveys a particular message, while strings B and C convey other messages. But, suppose I want to send you more than three different types of message. By pulling on multiple strings simultaneously, I can extend the number of possible messages to seven (A, B, C, AB, BC, AC, and ABC). This is the best I can do, without using something like a 'Morse code' style of temporal pulling pattern. But the idea of using multiple strings is actually very expressive for larger numbers of strings. Ten strings can convey about a thousand messages, and twenty strings can convey about a million messages. A thousand strings can convey about 10^{300} messages (for comparison, there are only about 10^{80} particles in the entire universe).

The firing of a neuron conveys a signal, like a quick tug on a string. And a neuron can convey a degree of certainty or urgency in the strength of its signal, sort of like tugging harder on a string. When a neuron's signal is strong, it fires more frequently. When a neuron's signal is weak, it fires less frequently. When two neurons are firing frequently, they are more likely to experience nearly coincidental firings. And, it is the coincident firings among many neurons that can express a wealth of information through an enormous range of patterns, like tugging on multiple strings at the same time.

Such is the relatively simple nature of a group of neurons. If the firing characteristics of neurons are indeed as simple as I've just described them, then any neural algorithm for processing the data conveyed by a group of neurons is likely to be pretty simple. Indeed, data can only be conveyed by the firing rates of single fibers and by the coincident firings across multiple fibers, and that's about it. There just aren't very many types of algorithms that can be applied to such simple types of signals. This leads us to believe that the algorithmic basis for cognition must be pretty simple, and that it achieves its tremendous power through massive redundancy.

There is another very powerful reason for suspecting that a simple neural algorithm is the basis for cognition, as Hawkins makes perfectly clear. The reason is revealed in a landmark paper by Vernon Mountcastle, a neuroscientist at Johns Hopkins University. The paper, titled "An Organizing Principle for Cerebral Function" (1978), clearly states that the neocortex – the outermost layer of the human brain believed to be responsible for human intelligence – is remarkably uniform in appearance and structure. Indeed, many brain researchers have since confirmed Mountcastle's finding that the cortex appears to have the same structure for processing auditory signals as for processing visual and, in fact, all types of sensory signals. On careful reflection, this should not surprise us, given the types of signals that all those cortical regions process. All types of sensory signals are just flurries of spikes occurring over bundles of fibers. In Hawkins' words:

> The regions of cortex that handle auditory input look like the regions that handle touch, which look like the regions that control muscles, which look like Broca's language area, which look like practically every other region of the cortex. Mountcastle suggests that since these regions all look the same, perhaps they are all actually performing the same basic operation! He proposes that the cortex uses the same computational tool to accomplish everything it does. ... The cortex does something universal that can be applied to any type of sensory or motor system.
>
> – **Jeff Hawkins**, *On Intelligence* (2004, p.50)

This is remarkable in its implication of underlying simplicity. There appears to be some relatively simple algorithm driving the self-organizing properties of the brain. That automatic organization indeed may arise from the previously described Hebbian principle for learning: Neurons that fire together wire together. It is a simple process that, over time and experience, can yield a very complex network of connections within the brain.

By recording coincident firings across multiple fibers, the brain encodes many statistical relationships between things sensed in the environment. It automatically becomes a flexible sort of simulator of the environment, by allowing the re-enactment of remembered relationships through what we refer to as the imagination. The re-enactment capability results from a sort of complement to Hebbian learning: Neurons that wire together become more likely to fire together. We'll learn in the next section exactly what enables the complement to Hebbian learning and how it allows neural firing patterns to serve as templates for their own replication.

Coincident firings between two or more sensory neurons can represent a real-world relationship of some sort. Such a relationship may be a very primitive one, but it turns out that relationships of any complexity can be built out of relationships of relationships of ever more primitive relationships. We may think of these relationships as being layered, one on top of another. Combining many primitive relationships into a complex hierarchy of relationships is a very important construct that allows the human mind to conceive of deeply abstract sorts of concepts, and to recognize critically important similarities between abstract concepts when such similarities exist. Let us contemplate how hierarchies of relationships automatically organize themselves.

Learning happens whenever two neurons at one level of a logical hierarchy coincidentally excite a common neuron at the next higher level. When the outputs of two firing neurons are able to excite a commonly connected third neuron, their connections with that common neuron are automatically strengthened, causing the common neuron to become more sensitive to simultaneous activity by the two coincidentally firing neurons in the future. This is how neurons that fire together, wire together. Let us examine this neural mechanism in more detail. We'll first need to briefly discuss the operational properties of neurons.

Human brain cells – neurons – typically have many inputs called *dendrites* and a single output called an *axon*. The dendritic structure is reminiscent of the roots of a tree, feeding input signals upward toward the axon, which metaphorically resembles the tree's long trunk. Dendrites serve as finger-like sensors, touching the axons of other neurons to sense when they are excited. When the dendrites of a particular neuron sense enough coincident activity for the sum to exceed a threshold, that event may cause the neuron to send an electrochemical *action potential* into its axon, thereby generating a spike of activity. It is then said to have 'fired'. The resulting spike, propagating along an axon, may then excite the dendrites of other connected neurons so as to cause *their* axons to fire.

The connection strengths between neurons are determined by *synapses* at the junctions between dendrites and axons. The physical characteristics of those connections get changed through the process of simultaneous firing. Synaptic connections are the malleable parts of the brain that become modified during learning. Collectively, they are precisely where a brain's information is stored. Somewhat similar to how a muscle that is flexed over and over again eventually builds up strength, individual synapses can likewise build up increased connectivity through repeated usage. Thus, the brain changes over time in two manners. Dendrites can grow into various positions of proximity with other neurons especially during childhood brain development, and the synaptic connection strengths between any two neurons can change as a result of the neurons simultaneously firing.

When two neurons that are connected by a synapse happen to simultaneously fire, the synaptic connection between them becomes strengthened so that the two neurons are even more likely to fire together in the future. Let's be a bit more specific: If the axon of neuron A is synaptically connected to a dendrite of neuron B, and if the firing of neuron A causes neuron B to fire simultaneously, then the synaptic connection strength will increase between them. If, in the future, neuron A happens to become excited enough to fire, it will in turn excite neuron B through the strengthened synapse. The increased connectivity between the two neurons makes neuron B more sensitive to stimulation by A.

A neuron typically needs to be excited by activity through several of its dendritic inputs before it will fire. After all, if neuron B were to fire every time neuron A fires, there would be no value in their master-slave relationship. But if neuron B were to be simultaneously excited by multiple neurons, call them A1 and A2, then B naturally becomes a coincidence detector. Neither A1 nor A2 can cause B to fire, but if A1 and A2 fire simultaneously, then neuron B fires and the respective synaptic connections both become strengthened. Simultaneity between A1 and A2 *causes* simultaneity between A1 and B, and between A2 and B. Neuron B thereby becomes more sensitive to nearly simultaneous firings between its inputs, A1 and A2. It effectively symbolizes a relationship of simultaneity between the two lower level neurons, which themselves may symbolize some sort of pattern or relationship sensed in the environment.

Now that we understand the basics of Hebbian learning at the neural level, there are a few more things we ought to understand about neurons, while we're on the subject. For instance, when we talk of connected neurons firing simultaneously, we need not expect the simultaneity to be precise. Simultaneous firings are those that occur within a small window of time. I won't try to define the size of that window; I merely intend to clarify that input firings need only be approximately simultaneous to cause output firings and subsequent synaptic strengthening.

A tiny and reliable delay between nearly simultaneous firings of neurons may indeed allow them to build sensitivity to motion. Neurons may in fact record sequence information between their inputs by encoding certain of them as 'arming' inputs, which must become active just prior to other inputs that then enable an 'armed' neuron to fire. For instance, if neurons A and B are connected to adjacent photo detectors in the eye, they may often fire in rapid succession in response to something moving through the field of vision. A higher level neuron to which they both connect may use that sequence information to detect and represent motion. This is as much as I wish to say about temporal sensitivity at the neural level, because I am already engaging in a significant degree of speculation.

In addition to the *excitatory* synaptic inputs to neurons, there are also *inhibitory* inputs as well. Whereas activity on excitatory inputs can *cause* a neuron to fire, activity on inhibitory inputs can *prevent* a neuron from firing. Inhibitory inputs come from special neurons that produce only an inhibitory neurotransmitter. Those types of neurons, whenever they fire, tend to inhibit all others with which they connect. There are many inhibitory neurons all throughout the cortex. We'll speculate on their specific role in later sections.

Finally, we need to understand that neurons have a *refractory period* after they fire, during which they are unlikely to fire again no matter what their inputs sense. It is as if a neuron gets tired after it fires and needs to build up energy before it can fire again. This short-term effect places an upper limit on how rapidly a neuron can fire. There is also a long-term fatiguing effect, known as *accommodation*, which further lowers the maximum rate of firing after a neuron has been heavily active for a while.

Firings may cascade through chains of connected neurons. And it is possible for a chain's output to be connected with a neuron earlier in the same chain, thus forming a loop of feedback. Firing patterns may in fact go around and around inside a given brain, sometimes shifting from one loop to another in the manner that a human's thoughts shift from one topic to another. We may speculate that, in the absence of other stimuli, the shifting can happen merely as a result of the natural fatiguing effects of heavily firing neurons.

Now, let us explore Hawkins' framework for the organization of cortical neurons. He describes the cortex – the outer-most layer of the brain – as a sheet of material, about the size of a dinner napkin, consisting of six somewhat distinct layers of cells. He asks us to imagine six business cards stacked up, as a visual model of the six layers in the cortex. If we were to examine a region of that cortex, we would find that it resembles a bundle of glued-together hairs. Imagine a thick round paint brush, whose fine bristles are glued together by dried up old paint, out of which we cut a cross-sectional slice to the thickness of six business cards. If the cross-section is laid flat on a table, the columnar bristles run vertically across the six layers of its thickness. While there are also many connecting fibers that run horizontally along the cortical sheet, information predominantly flows vertically, across the layers, within a column of neurons. Each column is considered to be a *cortical unit*.

Let us temporarily ignore the six layers and simply think of the cortex as a single layer sheet on which any spot corresponds to a single cortical unit. In this section of the book, I'll often treat the cortical unit as if it were a single neuron, having many dendritic inputs and one axonal output, even though it may more accurately be considered as a group of partnered neurons in a column that vertically spans the six distinct layers of the cortex. Just keep in mind that a cortical unit may include some added functionality, over and above what a single neuron can do. However, we won't need that added functionality for our immediate purposes.

Through far-reaching lateral connections, the output of one cortical unit can serve as the input to another cortical unit in a completely different region of the cortex. Indeed, there are sets of such lateral connections that seem to run in a parallel, or nearly parallel, fashion between various coin-sized cortical regions, linking the outputs from units in one region to the inputs of respective units in another region. By following an axon from one region to another, we find that when it reaches its destination region, it comes in contact with thousands of dendrites from immediately surrounding cortical units.

We may think of regions connected by far-reaching parallel axons as being logically stacked, one on top of another, as if the inputs of each region come from a level below and the outputs of each region go straight up to a level above. How high does the logical stacking go? I don't presume to know, and I can only answer by speculating that more levels in the stacking allow deeper and more abstract concepts to be understood. This hierarchical architecture is very similar to the hierarchical connectionism discussed in chapter five.

Stacks of cortical regions logically combine to form levels upon levels of cortical units, which we will soon visualize as automatically connecting themselves into many overlapping hierarchies, each representing a concept. There is a stack of cortical regions for each of the senses, but we may presume that all the independent sensory stacks begin to merge at higher levels, overlapping in ways that allow a high-level concept to relate all the relevant sensations. For example, the concept of a garden relates the various sights of plants, the sounds of birds and bees, and the smells of flowers. And, even within a single stack, connections between levels probably don't run strictly parallel, but rather, tend to fan-out and intermingle at higher levels.

As I speak of 'levels' in the cortex, I'll always be referring to the logical stacking that results from the outputs of one region being connected to the inputs of another. These levels are distinctly different from the six 'layers' that make up the cortex itself. The *layers* of the cortex give added functionality to the cortical unit, whereas the *levels* of logical stacking allow hierarchies of relationships to be expressed. I'll maintain the distinction by using the words *layers* and *levels* so that they always refer to the same respective things.

Let us consider an example of how the total system works. Imagine the retina on the back of an eye that is looking at some pattern, such as a cross. The image of the cross is projected by the eye's lens onto the photoreceptors in the retina. A corresponding pattern of photoreceptors will become excited in response to the pattern of projected light. The photoreceptors send their firing signals 'upward' to neurons in some region of the cortex, and some cortical neurons will fire as a result. Any cortical neurons that just happen to be well-connected to a bunch of firing photoreceptor axons will be more likely than others to fire. Those neurons 'fit' the pattern being viewed. Simultaneous firings between photoreceptor axons and responding cortical neurons will cause the associated synaptic connection strengths between them to increase. Long after the cross disappears from the field of view, the corresponding pattern of enhanced connections remains.

The enhanced connections in the cortex need not take the physical shape of a cross, but the mapping from the cortex to the retina will be such that the remembered pattern in the cortex corresponds to the original shape of the cross on the retina. If the cortical neurons that fired could send signals back through the connections that got strengthened – back toward the photoreceptors – the signals would go to photoreceptors in the original shape of the cross. The mapping between real-world shapes and the corresponding cortical connections is merely consistent, not necessarily topographically identical.

If cortical neurons were to have long enough dendrites and sufficient numbers of synapses to connect with every photoreceptor axon coming from the retina, then any given visual pattern on the retina could be represented by any single cortical neuron. But, the dendrites of each cortical neuron only connect with a small localized subset of the total number of photoreceptor axons. So, any unifying connections between first-level simultaneously firing neurons that happen to be widely separated must be made at higher levels in the logical stack of cortical regions.

Now, let us contemplate what happens as the outputs from the first-level cortical region arrive at the region corresponding to the second level in the logical stack. The very same sort of process happens in the second level as happened in the first level. But instead of learning patterns in the photoreceptor firings, the second level learns patterns expressed by the first level of cortical unit firings. Those are second-order patterns, or, patterns of patterns. The third level then learns patterns among the second level firings, and so on, up the stack of logically connected regions. This arrangement of Nth-order relationships builds hierarchies of connections from the bottom, up. Each hierarchy represents what we typically think of as a *concept*.

Let us imagine that, as widely distributed instances of *related* excitations climb higher in a stack of cortical levels, they can move laterally toward each other, converging at some high-level node, which symbolically represents an underlying relationship among the many active units that feed it. Such a symbolic node, in order to become active, may require excitation from many specific units below it, thereby symbolizing some sort of real-world relationship or pattern between them all. Neurons at high levels in a hierarchy represent abstract relationships in the real world. The higher the level, the more abstract the relationship. Each high-level neuron of a mature brain, representing some high-order relationship, can be considered as a module that monitors the environment for some particular abstract sort of pattern.

By way of these automatically occurring hierarchical relationships, there is a beautiful compression scheme at work in the cortex. We'll discover that, as the lowest level of visual neurons collects a library of small fundamental patterns that tend to reappear again and again in the visual field, the next-higher level concurrently looks for patterns composed of those patterns. By repeating this scheme over many levels, the resulting memory system takes the logical form of many overlapping hierarchies. The compression of encoding results from the fact that all the hierarchies tend to share the same detailed representations for primitive patterns represented at the lowest levels, but they become more differentiated at higher levels.

Allow me to illustrate the concept of Nth-order relationships through a simple example. Let us contemplate the simple concept of a *door*. A typical door is a relatively more complex entity than the simpler things of which it is composed. The simpler patterns include such things as hinges, a knob, a lock, a peephole, a rectangular wooden slab and a brass knocker. Those things come together to create a typical front door. In the case of a door, or any other sort of composite object, a mind must learn to recognize the lower-level component patterns before it can fully understand the more complex, aggregate pattern.

Once a mind recognizes a wide range of low-level things, those primitive patterns can be used in the representations of other complex objects. For instance, once a mind recognizes the pattern of a hinge, that same pattern can be used in the representation of other things having similar sorts of hinges. Conceptual hierarchies overlap heavily at lower levels because they tend to share primitive patterns. And, after all, they all share the very same photoreceptors as inputs. The higher-order patterns represented at higher levels of a conceptual hierarchy are more abstract, as they are more removed from the concreteness of sensory data.

Once the hierarchical recognition of objects is in place, the mind can then use those same hierarchies to make predictions about the world. Our minds make such predictions all the time without our consciously being aware of it, as Hawkins clearly illustrates in the following quote:

> Suppose while you are out, I sneak over to your home
> and change something about your door. ... I could change
> its color, add a knocker where the peephole used to be,

or add a window. I can imagine a thousand changes that could be made to your door unbeknownst to you. When you come home that day and attempt to open the door, you will quickly detect that something is wrong. The point is that you will notice any of a thousand changes in a very short period of time. ... There is only one way to interpret your reaction to the altered door: your brain makes low-level sensory predictions about what it expects to see, hear, and feel at every given moment, and does so in parallel. All regions of your neocortex are simultaneously trying to predict what their next experience will be. ... "Prediction" means that the neurons involved in sensing your door become active in advance of them actually receiving sensory input. When the sensory input does arrive, it is compared with what was expected. ... Prediction is not just one of the things your brain does. It is the *primary function* of the neocortex, and the foundation of intelligence.

– **Jeff Hawkins**, *On Intelligence* (2004, pp. 87-89)

Hawkins' example shows how our minds automatically take notice of things that seem wrong, or out of place, resulting from predictions of the present based on the past. But, even more importantly, we'll see that our minds can also automatically anticipate the future results of contemplated actions. Both of those predictive abilities would certainly have been of evolutionary advantage to our ancestors. And they continue to be of tremendous advantage to our modern selves. Acts of prediction enable us to build elaborate things with confidence that they will work as we expect them to. And, as previously discussed, prediction provides the rational foundation for the principle of investment in all types of complex cooperative ventures.

Hawkins refers to this neural architecture as the *memory-prediction framework*. As it turns out, prediction and memory are absolutely identical. Here's an illustrative example. If I ask you to imagine going to the dentist tomorrow, to get a cavity drilled and filled, you may feel a bit uneasy about it. My mere mention of going to the dentist causes you to predict what it will be like. Perhaps you can imagine the feeling of the drill hitting a nerve in your tooth. That imagined feeling of prediction will correspond directly to some similar feeling you've had at the dentist in the past. In fact, the whole imagined procedure of driving to the dentist, parking the car, going inside, notifying the receptionist and sitting in the dentist's chair will exactly mirror your memories of experiences you've had at the dentist in the past. You won't imagine going to my dentist, because you have no memory of my dentist. A brain's prediction of the future always mirrors its own memory of the past.

So then, how does prediction of the future happen? Is there an identifiable manner of signal flow that might be predictive in the hierarchically organized neural structure defined so far? Yes, such prediction could easily be explained by a downward flow of neural activity that flows in parallel paths but opposite directions to the upward flow. The upward flow corresponds to the creation of memory, and the downward flow corresponds to mirrored prediction.

Downward Signal Flow

Let us briefly review the model of neural architecture presented so far. As real-world patterns are sensed, they are transduced to neural firings, and are then propagated upward through more and more abstract levels of hierarchical relationships. But they only propagate upward so long as they fit familiar patterns represented by established neuronal connections at higher levels.

Now, we might suppose that there is in your brain a neuron (or set of neurons) at some relatively high level that abstractly represents the concept of a 'front door', for example. It becomes excited by the concurrent excitation of lower-level concepts, as happens by the recognition of a rectangular slab, a door knob, a brass knocker, a lock, some hinges, and so on. Your brain probably has many such cortical sites, each corresponding to a particular front door with which you are very familiar (your own, your friend's, your parents', etc.). Let us focus on your brain's representation of your own front door.

Every time you go through your own front door, the corresponding neural pathways are reinforced. They dutifully direct neural activity upward from sensory data to the conceptual high-level neuron, or set of neurons, corresponding to your front door. They do so simply because the patterns in the sensory data precisely match patterns with which you have become very familiar.

Now, suppose that, for every neural pathway up the hierarchy, there is a corresponding parallel pathway back down. And suppose that, whenever the high-level neural site corresponding to your front door is excited, such an event causes neural activity to propagate downward through all the paths that parallel the typically occurring upward paths. So, as you approach your own front door, you sense all the familiar objects that compose it, and, as a result, neural activity flows upward to the corresponding high-level site. Consequently, that site sends neural activity back down to all the low-level neural sites corresponding to the familiar components of your own front door. Any deviation from what you have historically experienced can become immediately obvious if we suppose that the brain has an automatic mechanism for comparing what it sees with what it expects to see. That is, at each cortical unit of a conceptual hierarchy, the upward flow of excitation is compared with the downward flow.

Recall that the cortical unit, consisting of six layers of cells arranged in a column, probably contains more functionality than that of a single neuron. The comparison between upward flow and downward flow may be part of that added functionality. But, are we justified in proposing a downward flow of activity that mirrors the typical upward flow? Are there empirical observations of such downward pointing neurons capable of providing the required feedback?

> ... there are as many if not more feedback connections in visual cortex as there are feedforward connections. For many years most scientists ignored these feedback connections. If your understanding of the brain focused on how the cortex took input, processed it, and then acted on it, you didn't need feedback. All you needed were feedforward connections leading from sensory to motor sections of the cortex. But when you begin to realize that the cortex's core function is to make predictions, then you have to put feedback into the model; the brain has to send information flowing back toward the region that first receives the inputs. Prediction requires a comparison between what is happening and what you expect to happen. What is actually happening flows up, and what you expect to happen flows down.

> **– Jeff Hawkins** (2004, p.104)

So, we may then speculate that the six layers of cells in each cortical unit, along with the feedback connections allowing signals to flow downward, may indeed provide for the extra degrees of functionality necessary to propagate and compare the upward and downward signal flows.

Perhaps upward and downward pointing neurons naturally pair-up and become linked together, as complementary pairs. Perhaps such a complementary pair can form something of a loop by way of Hebbian learning. Perhaps upward-pointing neurons learn which of the many nearby downward-pointing neurons are their complementary mates by the immediate downward feedback that occurs, in response to firing. Perhaps the first downward-pointing neuron to respond inhibits the firing of other nearby downward-pointing neurons. Perhaps conversely-pointing pairs become linked as mates through the nearly simultaneous firings that result from round-trip activity. We'll later discuss just how such reciprocal connections can be made exclusive through immediate inhibition of other local neural firings.

Let us proceed under the assumption that the mechanism of mirrored downward flow is responsible for the *prediction* part of Hawkins' *memory-prediction* model. That mechanism enables the complement of Hebbian learning to which I previously referred: Neurons that wire together tend to fire together. By supposing that the downward flow of neural activity can then re-excite parallel upward-flowing activity, we may thereby imagine an automatic re-enactment of familiar firing patterns in a way that allows us to predict what should be happening, or what is likely to happen, in a given circumstance.

Mirrored downward flow is the mechanism by which common patterns of neural activity are guaranteed to serve as templates for their own replication. By sending neural flow back down whatever hierarchies have been excited at the top, our brains automatically and constantly cause both an expectation of the present and a prediction of the future. Indeed, those downward flowing signals perform several very important functions. Let us consider them in turn.

First, as Hawkins clearly showed, the downward flowing neural patterns allow our brains to detect things that are out of the ordinary. Imagine seeing a face without a nose, or with a prominent scar. Your mind will automatically attend to the anomaly. How could it possibly do that unless it uses a standard model of a face for comparison. This mechanism of comparing incoming patterns against familiar patterns allows us to automatically discover all sorts of things that are unusual. Our ancestors would have surely gained evolutionary advantage by attending to things that were out of the ordinary. For example, a localized aberration in the gentle swaying of tall grass might have indicated the presence of a predator to our ancestors.

The **second** important function for the downward flowing neural patterns is to fill in whatever gaps exist in impoverished sensory data. For example, when a football player looks out at the world through the bars across the face of his helmet, his brain automatically fills in the world behind the bars. If one of the bars cuts through the body of an oncoming tackler, such perceived bisection doesn't change the perceived danger of being tackled. The player's brain knows that the oncoming tackler is indeed an intact individual, even though the eyes see a broken image of the opponent. The perceived pieces of the opponent are enough sensory data to propagate upward and excite a set of neurons that represent the symbolic representation of a typical opponent player. That excited representation then causes a downward flow of information regarding what a prototypical opponent player looks like, in its entirety. At lower levels, closer to the senses, neurons that correspond to visually obstructed regions are excited from above according to previously remembered patterns. The partially obscured visual field is thus 'filled-in' from above using expectations that have been learned through experience.

The **third** function of downward flow is to implement imagination. As you read this, you may be far away from your front door, and yet, you can easily conjure up a mental image of it, complete with all its components. Just by thinking about your own front door, as prompted by what you are now reading, your brain excites the corresponding high-level neural site, which then sends activity downward to all lower-level component sites. Your imagination is able to focus in on any of the component parts of your own front door by sending downward flow along whatever pathways happen to run conversely parallel to the typical upward-flowing pathways that have been previously excited each time you have encountered your own front door. This lays the foundation for a dynamic imagination that not only conjures expectations of the present, but also, predictions of the future.

The **fourth** thing the downward flow of information does is to motivate muscle contractions, thereby causing bodily movement. For instance, when an interesting pattern, such as an attractive face, is recognized in the periphery of the visual field, the eyes are automatically instructed to shift their gaze in a way that brings the interesting pattern onto the highly concentrated region of photoreceptors at the center of the visual field (the fovea). Notice that the downward flow is directly *caused* by upward flow. There is no other source of volition that is able to initiate downward flow of information.

Let us suppose that all controllable muscle actions result from neural activity flowing downward through cortical levels, proceeding out of the brain and into the body's nervous system, where it eventually reaches the muscles themselves. For controlling muscles, the abstracting nature of the hierarchy again works to our benefit, but in the opposite direction of information flow, from top to bottom. As signals propagate downward, they become less abstract and more concrete – less symbolic and more elaborated.

The act of throwing a ball, for example, is an abstract concept that can start at a single, high-level node, but is translated into broader and broader sets of neurons at lower levels. The single concept translates into the many sub-actions of: cocking the arm, stepping forward, twisting the torso, flexing the elbow, flicking the wrist and releasing the fingers. Each of those sub-actions further translates into many different muscle contractions. At the lowest levels of neural output, there is a one-to-one correspondence between neurons and possible muscle contraction signals. At that low level, the very complicated muscle contraction sequence emerges from the brain and heads toward the respective muscles.

Whether or not the downward flow of neural activity results in muscular actions of the arms and legs depends on whether the mind is in a mode of execution or contemplation. So, there must be some sort of gating mechanism that releases the signals to muscles during a mode of execution.

The mental predictions of various muscular contraction sequences follow directly from previous observations of one's own actions. So, a baby learns to grasp when it randomly squeezes its fingers around some object and witnesses the resulting unity of hand and object. Any future grasping performed by the baby results from a mental prediction, and a consequent belief, that such an action will again cause a unity of hand and object. In essence, the baby's prediction mirrors its previous observation of its own actions.

As a baby observes itself acting in various random ways, it builds up a library of things it knows how to do. Here's how: Various randomly initiated downward flows cause various muscle contraction sequences, the effects of which are simultaneously observed. The upward-flowing afferent neurons of observation become causally linked to the downward-flowing efferent neurons of muscular control whenever they are simultaneously active. The causally connected, conversely pointing pathways thereby become useful for predicting the outcomes of various muscle contraction sequences when the mind is in a contemplative mode. Those connections allow our brains to contemplate and plan our actions well before we do them.

Thus, we have touched on the **fifth** and most important function of downward neural flow – its crucial participation in the dynamic representation of causality. Please pay close attention now, because I'm going to slowly reveal the very specific mechanism by which causal memes are stored in the brain. The mechanism is completely consistent with Hawkins' framework.

The Neural Representation of a Meme

The real value of any meme is in its dynamic nature. A static meme is pretty much worthless, unless it is able to be translated into some dynamic process. By thinking at the most fundamental level, it is apparent that *value* and *worth* are defined by evolutionary fitness, and fitness always depends on an ability to adapt, to change, to move, and to do. Evolution is a dynamic process and so is life. In fact, the definition of life that I earlier laid out requires some sort of pattern that directs available energy toward its own replication. An ability to direct energy necessarily involves an activity or behavior. And so, instead of focusing on static patterns themselves, we must now shift our focus to the dynamic events of replicating and expressing patterns of behavior.

Regarding both genes and memes, it is the replicated pattern of activity that is most important, not the static pattern that descriptively symbolizes the activity. Indeed, a gene is worthless without the mechanism for duplicating it or expressing it as a phenotype. So, as we contemplate a model of the human mind, its most important feature will relate to how it encodes dynamic characteristics – sequences of activity, motion, behavior and common progressions. But the hierarchies we have so far discussed represent static patterns, not dynamic ones. So, how does the human brain become a movie-camera, as opposed to a still camera? That is what we hope to answer.

Imagine walking up to your own front door with a key in your hand. What happens next? You automatically imagine inserting the key into the lock and turning it. What then? You imagine turning the door handle. And then? You imagine the door opening. Which way does the door swing – into the house or out of it? I'm sure I don't need to answer that – your brain has already imagined it, automatically.

Causal linkages between various conceptual hierarchies automatically guide the flows of our imaginations. Those linkages represent the dynamic nature of the world, and they are automatically acquired through all our various experiences. We may roughly analogize each conceptual hierarchy as a static frame in a movie, and the causal linkages between them as the sequential ordering of the frames. Unlike a movie, however, the many sequential linkage pathways in a brain can converge, diverge and intersect at various points. They are something like the behavioral 'scripts' that we previously discussed.

Let us recast our previous discussion of intersecting scripts in terms of intersecting strings of causal memes. We'll do so by analogizing a causal meme as any remembered sequence of movie frames. The first frame in a sequence defines the input state of a causal meme, and the last frame in the sequence defines the output state of the meme. The change between the input and output states represents the causality. Keep in mind that each frame of the sequence corresponds to a cortical hierarchy, and that each hierarchy has both an upward and a downward-flowing side. Inputs correspond to upward-flowing hierarchies, and outputs correspond to downward-flowing hierarchies.

Now, let us imagine that the brain maps all its visual observations, analogous to movie frames, into a conceptual space such that very similar frames map to roughly the same point in that space. So, for example, every remembered 'movie' (experience) that has a scene involving a person standing in front of a door with a key will intersect at a point in conceptual space corresponding to that particular abstract concept. The entire system, then, analogizes to a tangled mess of movie filmstrips intersecting at many points where their frames are conceptually similar. Now, with this background analogy in mind, we may contemplate the functional capabilities of dynamically linked hierarchies, as if they are conceptual frames in something of a movie.

The similarity of brain encoding to a movie enables the brain to know about the future of things it encodes and subsequently recognizes, simply by playing whichever strip of film best matches the current situation. Given a particular brand new situation, the brain maps the corresponding patterns into its vast collection of movie snippets. If the new situation closely maps to some existing frame of a remembered movie snippet, then the brain simply 'replays' that best-matching snippet in order to predict the likely future. So, the act of understanding a dynamic situation amounts to finding a frame, among the millions of movie clips in one's brain, that closely matches the contemplated situation.

When seeing common activities performed, visual patterns are typically followed by other visual patterns, usually in a similar pattern of temporal order, like the frames of a movie. So, when a child witnesses its parent unlocking the door to the family home, the pattern of turning the key is always followed by the pattern of turning the door handle. In the child's mind, the concept of 'unlocking' becomes linked to some new state (unlocked) that permits the actions of 'turning the door handle' and 'opening the door'.

There is a conceptual hierarchy for each of many differentiated states – for each unique frame – in every movie-like experience one has ever mentally encoded. Like the frames of a movie, successively occurring hierarchies are causally linked. For instance, there is a high-level concept of 'unlocking a door', which involves several causally linked sub-hierarchies that we might refer to as 'inserting a key', 'turning the key', 'turning the door knob' and 'opening the door'. The visual perception of both a key and of a door lock satisfy the required upward-flowing condition for the fulfillment of the initial phase of 'unlocking a door'. That is, it satisfies the input pattern of the sub-hierarchy corresponding to 'inserting a key'.

In addition to causal linkages between hierarchies at high levels, there are also linkages at lower levels. These lower level linkages allow us to do two important things: they enable us to imagine plans in various levels of detail, and they enable us to imagine switching from one causal filmstrip to another at points where they conceptually intersect. Thus, we can contemplate various options and alternatives within our plans. Let us consider an example that illustrates how these capabilities are possible.

Whether the specific act of inserting a key is actually perceived or only imagined, the corresponding hierarchy becomes active, including both upward and downward-flowing components. The downward-flowing hierarchy corresponding to 'inserting a key' creates low-level firings that satisfy the required upward-flowing condition for activating the upward-flowing (input) hierarchy corresponding to the act of 'turning the key'. Activation of that high-level concept causes downward-flowing neural activity, which then satisfies the required upward-flowing condition for the act of 'turning the door knob'. The concept of 'turning the door knob' sends activity downward so as to predict the lower-level patterns corresponding to the holding of a door knob attached to a door that is completely unlatched. Such a low-level pattern is sufficient to excite the upward-flowing input hierarchy corresponding to the concept of 'opening the door'.

Downward flows, from sequentially linked concepts, provide imagined conditions that satisfy the input patterns of other sets of upward flows. Now, realize that it may sometimes take multiple hierarchies of downward flowing activity to satisfy the upward patterns corresponding to a given concept. For instance, the concept of unlocking a door requires the simultaneous activation of the respective downward-flowing hierarchies for both a locked door *and* a key. Only the combination of those two output hierarchies can satisfy the more complex input hierarchy for unlocking a door.

So, while high-level linkages connect static hierarchies into scripts, as if they are like frames in a movie, low-level linkages can allow the imagination to travel up and down various connected hierarchies, possibly switching from one filmstrip to another at points where they intersect in conceptual space, or at points where multiple outputs sufficiently satisfy a single input. These effects will become more clear when we later discuss complementary interfaces and inductive similarities between hierarchically encoded patterns. Let us maintain our current discussion at a fairly abstract level.

The most important function of downward flow should now be apparent. It provides linkage, within the imagination, between various sets of complementary activities that may be performed in meaningful combinations and sequences, but that haven't yet become linked at high levels through experience. It provides pathways for neural activity flows that enable our imaginations to be dynamic, exploratory and creative.

Neural activity flows up and down cortical hierarchies that are linked at high levels by experienced causality, and at low levels by their mappings to primitives of shared commonality. These roundabout flows cause us to automatically imagine all sorts of combinations of complementary activities. The natural flows of our thoughts, during quiescent moments, follow these pathways, up and down the cortical hierarchies. They are constantly causing us to imagine possible future scenarios. They are constantly predicting what we should expect to happen next, given the circumstances, and they guide our motor behaviors according to what we expect is appropriate. Here is how Hawkins' describes the situation.

> With humans the cortex has taken over most of our motor behavior. Instead of just making predictions based on the behavior of the old brain, the human neocortex directs behavior to satisfy its predictions. ... It is constantly predicting what you will see, hear, and feel, mostly in ways you are unconscious of. These predictions are our thoughts, and, when combined with sensory input, they are our perceptions.
>
> **– Jeff Hawkins** (2004, p.104)

A round trip up the input of one cortical hierarchy and back down the output of another sequentially linked hierarchy is representative of a causal meme. The upward hierarchy determines the input pattern, and the downward hierarchy determines the output pattern. Recall my previous analogy for constructing plans by stringing together causal memes as if we were stringing together dominos by matching the patterns of dots on their ends. Now we can see exactly what types of multi-dimensional patterns define the complex inputs and outputs of causal memes. They are hierarchically nested relationships of familiar real-world patterns.

Notice that sequentially linked hierarchies may be traversed at any level of granularity, or detail. Going up and down at a low level corresponds to the examination of an elaborate plan in extremely fine detail, considering every individual step in the plan. On the other hand, following sequential linkages at a high level corresponds to conceptualizing a plan without considering the intricate details.

For two causal memes having complementary interfaces, their matching input and output hierarchies will already share many lower-level nodes. In fact, such significant low-level sharing is what makes them complementary. This means that the appropriate connections between all complementary causal memes already exist in a mature brain having lots of knowledge, by virtue of shared nodes. To imagine a plan, a brain need only identify and activate any of the many existing crisscrossing strings of causal memes. At a point of intersection where the causal linkages are able to equally excite two or more paths, alternatives exist, and a decision needs to be made. We'll later see that decisions are automatically made by an evolutionary mental process of trial and error, exciting various causally connected pathways, by imagining the performance of them, to see which is best. Whichever path ultimately leads to the greatest amount of expected happiness is automatically chosen as the path to execute.

Once the cortex predicts a clear path to lots of happiness, it merely executes that path. As Hawkins says, "the human neocortex directs behavior to satisfy its predictions." If you want to see a funny example of this phenomenon, find that one spot on your dog or cat, near its neck, that, when scratched, causes the animal to move its hind leg as though its leg is what is doing the scratching. On the basis of your scratching, the animal's brain has apparently predicted that future scratching of the same spot would feel good, so it directs its hind leg so as to satisfy its prediction.

Once an imagined plan has been adopted for action, the execution of that plan merely follows the same neural pathways as the prediction did. But the downward signals continue their downward treks right on through intermediate nodes corresponding to abstract representations of behaviors, down toward the motor system, branching into ever more detail, proceeding out of the brain, eventually reaching the appropriate muscles. Just as Hawkins suggests, we automatically act so as to fulfill our preferred predictions.

Resonating Circuits

We have seen how a unit of causality is captured by a linkage between conceptual hierarchies as a result of exposure to some common causal sequence. Because of such linkage, neural activity can go up one hierarchy, and come down another that is linked, thereby predicting the consequence of causality. While this sort of path is critical for generating expectations and visualizing the dynamics of causation, there is another route that neural activity might take. It is in fact strictly circular, going up and down the very same hierarchy. It is just this sort of circular resonance that might account for the temporary persistence of short-term memory.

As sensory activity enters the cortex, it excites neurons in the lower levels of whatever remembered hierarchies happen to exist. If the pattern of incoming sensory activity matches a pattern encoded by one of the hierarchies, the activity will proceed to higher levels by way of neurons that I'll refer to as *first-responders*. Once a high-level node is reached, it will send activity back down a parallel hierarchy of the same structure so as to fill-in any impoverished sensory data and make predictions about what is being sensed.

Now, let us imagine that, intimately associated with each first-responder neuron, there is a group of other 'partner' neurons that work together, in conjunction with the first-responder, so as to give greater functionality to the corresponding hierarchical node. Recall that we are justified in postulating such partnered neurons on the basis of the six layers of cells in a cortical column, which apparently constitutes a cortical unit – a hierarchical node. Indeed, a cortical unit, being composed of multiple neurons, just might be capable of doing some things that we don't normally associate with individual neurons.

We have already discussed the possibility of nodes performing comparisons between upward flowing sensations and downward flowing expectations. Let us now speculate on a third neural pathway back up again. We'll call these *re-responders*. If downward flowing feedback is able to excite these upward-going re-responders, and the re-responders are able to re-excite the downward flowing feedback from the level above, then we can imagine a resonant loop that 'rings' for some time after the first-responders stop firing. This 'ringing' enables a short-term memory of parallel circuits resonating in response to first-responder circuits that were active a little while ago.

I don't see how first-responder circuits can participate in the resonating loops of short-term memory, because if they did, it would be impossible for us to mentally differentiate between experiences of the recent past and of the immediate present. There has to be a difference between neural activity that represents the recent past and that which represents the present. So, let us assume that first-responders represent what is being currently sensed, and that a secondary set of pathways implements the resonance of short-term memory. There may in fact be several 'spirals' in the set of pathways that constitute a resonating short-term memory loop. And some of those spiral connections may be shared among several similar hierarchies. Such an architecture might allow our thoughts to automatically shift among similar variations on a conceptual theme.

I am portraying circular circuits as being implemented within the cortex, but it is quite possible that resonant persistence occurs by way of circular connections between the cortex and other brain systems, instead. For example, the thalamus is known to have many reciprocal connections with the cortex, forming loops of recurrent activity. I am really only interested in establishing some logical neural mechanism of short-term memory that allows the concepts of our thoughts to remain within the focus of attentive consciousness for a short period of time. Resonating loops seem to be the best candidates for short-term memory.

Just as the ringing of a bell continues for some time after it is struck, resonant circuits continue their activity well after the first-responders have ceased firing. Such a mechanism of resonance, yielding a short-term memory of events, allows the cortex to be sensitive to associations between patterns even though their occurrences may be spread out over a little bit of time. So long as two patterns occur within a resonance period, they can cause simultaneous firings to establish linkages between their associated hierarchies. And, since linkages can occur between the first-responders of one pattern and the re-responders of another pattern, such linkages carry an implication of temporal ordering.

The shadowing mechanism of resonance also allows a distinction to be made between real and imagined sensations. First-responders are real sensations, whereas re-responders are imaginary and may be stimulated merely by patterns of thought. Since re-responders represent exactly the same relational encoding patterns as first-responders, they can accurately recreate memories and thereby simulate previously experienced reality. Indeed, the flows of neural activity that result from imagination will tend to excite hierarchies in whatever sequence they naturally tend to occur. Our brains, in essence, automatically become simulators of the environment. So, in addition to representing our memories, these re-responder circuits might also represent our imagined expectations.

Let us suppose that re-responder circuits become active whenever we consider things from the past, and also, whenever we think of what might happen in the future. When we contemplate going to the dentist in the future, we are exciting the very same circuits that become excited when we think of our past experiences at the dentist. Those circuits are the very same circuits that were resonating with activity during the moments just after our actual experiences at the dentist. We now have a model of cognition that accounts for much of what we mentally experience.

In a later chapter I shall use our current model to show how the fatiguing effects of neurons can cause resonating loops of activity to 'search' many existing causal pathways for the best one. As neurons in a loop get fatigued from firing, the resonant activity will shift to similar looping pathways, thereby exciting similar conceptual hierarchies. We'll later see how such mutations, applied to patterns of neural firings, allow a mind to consider only relevant variations on a theme. The scheme allows a better-than-random style of thought to creatively evolve (by Lamarckian means) through a meiotic-like mechanism of meme recombination. It is a more efficient style of evolution than the random plodding of biological evolution. But it is a style of evolution nonetheless.

Nth-Order Relationships

Let us now return our attention to the upward-flowing neural activity, ascending through levels of first-responder neurons. At each level is a huge collection of pattern detectors that automatically organize themselves so as to recognize commonly occurring patterns coming from levels below. By putting pattern detectors on top of pattern detectors, the brain achieves more and more sophistication in the abstract types of patterns it can recognize. We'll also find that nested pattern detectors provide a very efficient compression scheme for data storage and a very important means for assessing a metric of similarity between any two patterns. But our immediate endeavor is merely to understand how nested relationships work in greater detail. So, let us consider how they relate to vision.

Neurons at the lowest level of the visual cortex, connected directly to visual photoreceptors, detect only small patterns, consisting of not much more than tiny fragments of edges at all orientations. Many edge detectors, spread over all locations of the retina, are then combined at the next higher level to form detectors of slightly bigger patterns, such as longer edges, corners and arbitrary curves. Those detectors are then combined at a yet higher level to form detectors for all sorts of complex patterns of shapes. And pattern detectors of complex shapes are then combined to form detectors of real objects. So, four or five levels of pattern detection can abstract from the minute details sensed at retinal photoreceptors all the way up to real objects.

We need even more levels of pattern detection if we are going to perceive how objects relate to each other. For example, the many components of a car relate to each other in the specific pattern of its formation. And, the sight of a car relates to the sounds and the smells that it makes. And big, loud, smelly cars relate to humans through patterns of usage. For instance, various patterns of driving a car, such as turning right and then left, can become integral components in even bigger patterns, such as going to the store. And the act of going to the store may be part of a higher-level plan to buy a tool for fixing the house. All plans are simply patterns of activity that are expected to yield benefit. From the simplest patterns of light, sound and smell that stimulate our senses, all the way up to the best laid plans of mice and men, all interesting aspects of the world are describable, hierarchically, in terms of patterns of patterns of patterns, and so on.

Every node in a conceptual hierarchy symbolically represents some relationship between events at lower levels. Nodes at the lowest level represent relationships between various features of the sensory data. But, let us be clear as to what we mean by *relationship*. As far as the brain is concerned, a relationship can only mean a nearly coincident firing between two or more neurons. A critical assumption, then, is that everything in the world is understandable on the basis of coincident activities. For instance, we teach our children language by simultaneously pointing to a book and saying the word 'book'.

While I can think of many primitive examples supporting the idea that a mind does nothing but detect coincident occurrences between patterns, there are, however, some modern concepts – much more complicated concepts involving relationships that don't occur simultaneously – that would have been impossible to convey before language. Perhaps language is required to conceptualize related events that are widely separated in time.

Consider, for instance, an alarm clock. How could you possibly convey the function of an alarm clock to someone who has never seen a clock before and who can't understand your language? There is no way, without words, to express the critical relationship that exists between the act of setting the alarm, now, and its eventual ringing. But with language it is easy: "This will ring a bell at sunrise." The words 'bell' and 'ring' conjure a coincident relationship in the mind that identifies a particular existing concept. Indeed, I automatically imagine a bell ringing, just from its mere suggestion. Next, the simultaneous consideration of a ringing bell and the word 'sunrise' coincidentally excites (in my imagination) their related hierarchies, just as if the sun were now rising and the alarm bell were simultaneously ringing.

A sentence is able to simultaneously excite certain hierarchies of a brain in a way that spans time and space. Indeed, it seems that language enables us to think deeper thoughts by allowing us to conceptualize such things that are causally related but that never occur simultaneously. Now, let us return to the easier study of related events that naturally occur at the same time.

It is easy to see how simple visual patterns are captured by coincident neural firings just after the patterns are transduced by the senses. But, what about such abstract relationships as, say, *love*? How does the brain come to recognize the relationship between two people who are in love? According to our model, such an abstract concept is necessarily built out of many slightly less abstract concepts. For instance, you would quickly detect the relationship of love if I were to show you a picture in which two people of similar age and opposite gender are smiling face to face, holding hands or kissing. You can detect all those features simply by patterns of patterns in the picture. The patterns representing faces, hands, and lips are merely related through higher-level patterns describing how those objects are relatively positioned.

You might even suspect the very abstract relationship of love after hearing a story of two people who constantly buy each other gifts, have common interests and always come to each other's defense. These concepts are all slightly less abstract than the concept of love, but are ultimately built from examples of concrete patterns. The fact that I can express each of them in short movie snippets demonstrates that they are all reducible to nearly coincident relationships. We are left to suspect, then, that every concept is able to be represented by an Nth-order relationship – a hierarchy of relationships N levels deep, having its 'fingers' on patterns of concrete sensory data.

We might wonder how big N needs to be to capture the depth of knowledge held by a typical human. In other words, how many levels are there in our most complex cortical hierarchies? I don't presume to know, although, I believe the game of *Twenty Questions* serves to illustrate the compounding power of Nth-order relationships. The game is played by having one person think of some object, any object, and having another person ask up to twenty questions in order to determine what the object is. It is usually pretty easy to pare down the entire world of objects, within twenty questions, to a single, particular object. Of course, thoughts involve more than just objects, but you get the idea.

The brain's ability to re-use primitive patterns in the recognition of complex patterns is extremely efficient. Not only is it efficient in the vast number of things it can mentally represent, but also in the ways it can easily learn brand new concepts that are built mostly from primitive concepts with which it is already familiar. Once our visual systems learn to detect edges and shadings of various orientations, higher levels of the visual cortex can then use relational combinations of those primitive features to describe and detect all sorts of common shapes, such as cylinders, spheres and planes. Once those common shapes are learned, they can be combined with perceptions of characteristics like texture and color into many sorts of naturally occurring complex shapes, such as tree trunks, berries and leaves.

The architecture, as we've described it so far, appears to be specifically designed to relate things that occur coincidentally. Additionally, it can automatically recognize some classes of abstract relationships that don't involve coincident occurrences, but that we intuitively know are important. For example, two things can become related, not by occurring simultaneously, but by always occurring in similar contexts. By this important mechanism we can build up classes of things that are abstractly similar in some regard. Let us consider an example, continuing a theme with which we are already familiar.

Suppose you have only ever seen one front door in your life – your own door. You've seen it many times after walking down your own walkway. The associated hierarchy for your door becomes active whenever you see it, and will likely be predicted merely from the act of walking down your own walkway. Now, suppose you were to visit a brand new house for the first time. As you walk down its walkway, which happens to be similar to your own walkway, your brain will automatically recognize the walkway as being similar to your own, and will thereby predict a door at the end of the walkway. It will predict your own door, because that is the only door you have ever seen. That prediction activates the hierarchy for your own door. But the new door, being somewhat different from yours, will cause the creation of its own hierarchy. Since the hierarchies for both your own door and for the new door are both active, they become linked at a higher level. That higher-level node will eventually come to represent the class of all front doors. The next time you walk down a new walkway for the first time, your brain will predict a prototypical door at the end of it that is some sort of combination of both doors you have now seen.

In this manner, the brain builds up classes of objects that tend to occur under the same or similar circumstances. This phenomenon of induction will be important to a later discussion of processing language, regarding the ability to automatically classify words as nouns, verbs, adjectives, and so on.

Specific doors are linked to a general class of all doors, and specific hands are linked to a general class of all hands, and specific keys are linked to a general class of all keys. Indeed, the act of unlocking a door will eventually become linked to the general classes of all doors, all hands, and all keys. And thus, the high-level neuron representing the concept of unlocking a door can indeed become active whether it is your door or my door, whether it is your hand or my hand, your key or my key. Abstraction and symbolic substitution seem to naturally emerge from hierarchical relationships.

The evolutionary development of symbolic substitution appears to have been critical to the development of mimicry. And the hierarchical representation of knowledge appears to be just the sort of structure required for symbolizing collections of objects and actions. Indeed, every node in a hierarchy symbolizes the collection of nodes beneath it. It seems that the ability to represent Nth-order relationships is a natural prerequisite for the evolutionary development of completely generalized mimicry. Consider a specific case of generalized mimicry.

When a child watches its father perform some sort of cultural habit in a given environmental situation, the child's mind remembers those patterns of behavior. The patterns are automatically recalled whenever similar environmental situations are encountered. If the child sees the parent starting to perform the same sequence of patterned behaviors again, in a later instance, the child's mind will automatically predict subsequent stages in the sequence of behaviors. All that is needed for mimicry to occur is the symbolic substitution of the child's self in place of the father. So, mimicry necessarily entails a notion of 'self'.

Symbolic substitution requires that patterns of behavior are abstractly represented in what Hawkins calls an *invariant* format. I shall now endeavor to show how such a critical phenomenon of abstract invariance comes about as a natural by-product of the hierarchical levels of Nth-order relationships.

Abstract Invariance

Abstraction is a difficult concept to convey. I'll be using the term 'abstract' to refer to concepts that are far removed from the reality that we can immediately see, hear, touch, smell, and taste. A good example is the weirdness of quantum mechanics. The more abstract a concept is, the harder it is to understand. Indeed, the very notion of abstraction is itself extremely abstract. Be that as it may, let's delve in.

Our brains are capable of extremely abstract thinking, thanks to the high-order relationships encoded by the hierarchical ordering of cortical neurons. The abstractness of a concept is related to its 'distance' (in hierarchical levels) from the measurable reality that we humans typically sense. A good example of an abstract concept is *praying*. It comes in various flavors of abstraction. For instance, a child's mind simply relates the concept of praying to the simple act of putting hands together and reciting some wishes. Such a concept is not very far removed from reality. It can easily be expressed in a picture. But, an adult, on the other hand, additionally considers the notion of God, the style of morality that God prefers we practice, and the idea that God doesn't like purely self-serving requests. So, the adult concept of praying can be a very abstract notion built of many other concepts, which are nearly as abstract, including the interplay between morality and the serving of one's self. Highly abstract concepts are difficult to convey in a picture, or even a movie.

Philosophers and scientists spend their lives trying to relate abstract concepts to real-world characteristics. Indeed, I spent a whole book relating the abstract concept of morality to the empirical evidence of evolution. In a deterministic universe, there is a reason for every event, and a relationship between everything and everything else. While all Nth-order relationships ultimately have their roots in the real world, the connecting pathways from high-order relationships down to those real roots may be very difficult to identify. Take, for example, the visual image of, say, a human face. What are the salient abstract characteristics that make a given face so distinguishable from other faces? That question is often explored from the top, down. But such an approach is unlikely to work. Visual relationships are built from the bottom, up.

There is no way to identify the high-level concepts that define the appearance of a face without first identifying the lower-level components from which its visualization is collectively composed. To understand a given level, we first need to understand the underlying level. The regress goes all the way down to the primitive visual elements at the lowest level. Once we understand the bottom level, we can build up. We'll soon speculate that the most primitive visual elements are edges, solids, shadings and textures, but we won't really know what the entire library of visual primitives actually are until we build a coincidence-detecting visual system and expose it to real scenery. All we can really know, without building a device and trying it, is that with each higher level we get closer to understanding the visually distinguishing characteristics of a human face. But, characteristics get more and more abstract in the ascension to higher and higher levels of a relational hierarchy. And, the only way we can understand abstract characteristics is by their complex relationships to reality.

Let us now move the discussion of abstraction and its consequent invariance to a slightly more technical platform, by means of a mathematical analogy. I'm afraid the discussion will necessarily involve a bit of hand-waving, but I believe there is technical merit here, waiting to be mathematically revealed. My familiarity with digital signal processing, which results from my background as an electrical engineer, enables me to see a great deal of similarity between the organization of the cortex and the organization of calculations required for performing a *Fourier Transform*. The Fourier Transform likewise converts information from a concrete, variant format to an abstract, invariant format. Allow me to share those insights as a means for illustrating how I believe the cortical hierarchy achieves its abstract invariance.

First, we need a little background regarding the types of patterns on which the Fourier Transform is most often used: Sound waves are fronts of air pressure that propagate through the atmosphere. They can be transduced by a microphone to an electric potential that varies with time. So, any given sound can be expressed as a function of time. One such function having special mathematical significance is the *sine wave*. It is a very pure sounding function. In fact, the sine wave function can be considered as a fundamental primitive out of which all sounds are built. By mixing together various sine waves of particular frequencies, phases and amplitudes, we can construct any imaginable sound. Or, conversely, we can decompose any given sound into its constituent sine waves.

The Fourier Transform is a mathematical formula for doing just that. It decomposes any complex sound wave into sine waves of various frequencies, phases and amplitudes. Here's an example: If, while sitting at a piano, you press and hold several keys simultaneously, the resulting sound wave will be a complex function of time, wiggling wildly, almost chaotically, up and down. Depending on which keys you press, and how well tuned the piano is, there may or may not be visually detectable repetition in the pattern of the wave. No matter. The Fourier Transform takes as input any complex time wave and produces as output an indication of its constituent frequencies. Those frequencies roughly correspond to the piano keys that were pressed. Even though the pattern of the sound wave varies wildly over time, its Fourier Transform remains invariant, correctly indicating that the sound remains similarly invariant, because the keys are invariantly being held down.

The Fourier Transform automatically picks out whatever sinusoidal patterns exist in a signal and thereby characterizes the signal in terms of those invariant patterns. It is a theoretical formula that is very difficult to compute, except in closed form for certain types of well-defined input signals. Luckily, there is a digital approximation, called the *Discrete Fourier Transform* (DFT), that is easily computable, even in real time (for low bandwidth signals).

Now, here is a coincidence that I happened to notice. The architecture of signal flow for the efficient calculation of the DFT looks just like the multiple overlapping hierarchies proposed in Hawkins' model of the cortex – one hierarchy per high-level concept, each being indirectly connected to, and dependent on, every low-level transducing element within its region.

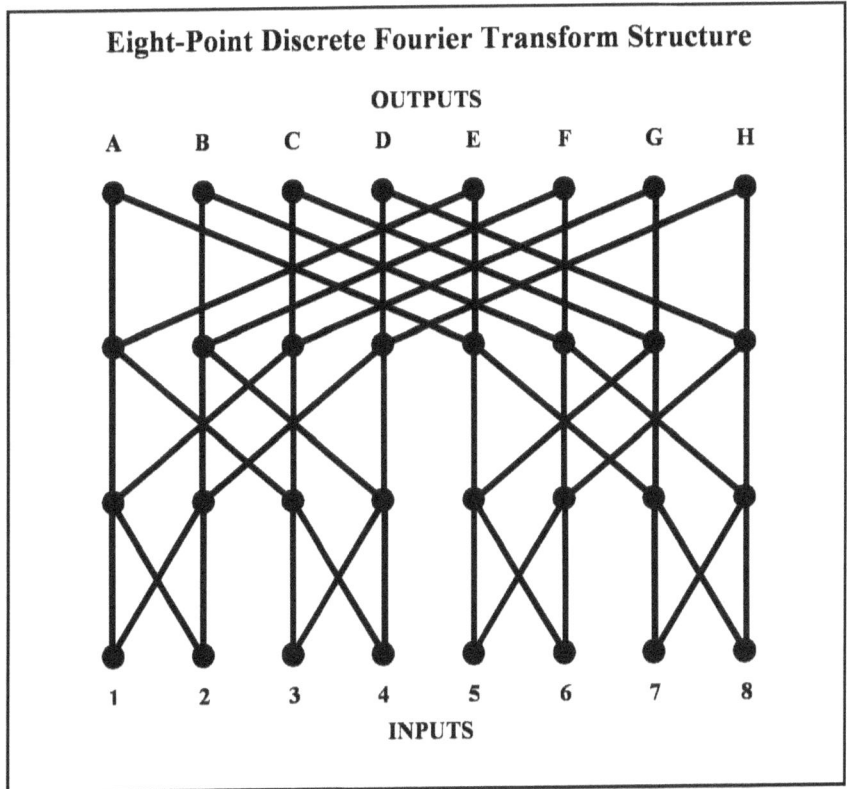

Eight-Point Discrete Fourier Transform Structure

OUTPUTS

A B C D E F G H

1 2 3 4 5 6 7 8

INPUTS

Figure 1

Figure 1 shows the structure of signal flow for computationally carrying out an eight-point DFT. By following the branches from each of the outputs, 'A' though 'H', it is easy to see that each one represents a four-level hierarchy, with a branching factor of two at each node, producing roots that touch all eight inputs. There is a great deal of similarity between this structure and the structure of a cortical hierarchy, as we've defined it. Of course, cortical nodes tend to have much bigger branching factors, more like a thousand connections per node. But the abstract relational nature of the structure is the same for both.

Why is this important? Well, it is crucial for illustrating a profound underlying transformation from the wild and woolly real world to some virtual domain of stable pattern representation. This each-to-every sort of architecture is capable of transforming from a domain of complex data to a domain of invariant patterns. It is something like the way a lens can gather and redirect many wildly divergent light rays so as to reconstruct the pattern of an image, or the way a simple recursive formula generates the beautifully ordered complexity of an entire fractal pattern. It is a mapping between the complex 'space' of reality and a much simpler 'space' in which the dimensions are simple patterns.

Just as the DFT transforms sound waves from the wildly varying time domain to the invariant frequency domain by picking out patterns based on sine waves, so does the human cortex transform the wildly varying sensory data from the real domain of space and time to some abstract domain of invariant representation, composed of relational patterns that are conducive to thought. They are the Nth-order relationships automatically constructed by Hebbian learning.

Returning from the esoteric arena of computational transforms, I'll now try to relate this concept of abstract invariance to a previous discussion of symbolic substitution. The previous example of symbolic substitution that I gave, related to the concept of unlocking a door, is certainly a style of abstract invariance. There are many others that naturally grow out of Nth-order hierarchies of relationships. Let us consider another style of example.

No matter in which musical key I first hear a tune, I can recite the melody in absolutely any key. As the tune goes through a form of hierarchical processing in my brain, similar to a Fourier Transform, it gets encoded in my memory in an abstract manner that is invariant of any key. When I recall the melody from memory, it is then translated from the invariant form to a form that is key-dependent. The abstract invariant format is simply encoded in terms of its patterns of intervals relative to the key. Unfortunately, I can't go into detail as to how the brain encodes music without explaining a bunch of intricacies regarding the cochlea – the mechanism of the inner ear. So, I'll go back to the domain of vision, and use another example: the simple recognition of any arbitrary rectangle independent of size.

The dynamic nature of memory plays a big part in the implementation of abstract invariance. Recall my suggestion that our brains are more like movie cameras than still cameras. Our memories don't store objects so much as they store sequences of actions. And so, there is likely no coordinate-based model of an actual rectangle in anyone's memory. But then, how is a rectangle recognized and understood? The answer has to do with how we perceive things. Suppose I show you a huge piece of paper with a very large rectangle on it, your eyes will first focus on one salient aspect of it, say, the upper left corner. There will be an automatic saccade (eye movement) along the top line, over to the upper right corner. Further such saccades trace the entire outline of the object. So, the brain sees, not a rectangle, but a sequence composed of various corners and straight lines. That dynamic sequence is a valid description for *any* rectangle independent of its size. It is abstractly invariant across all large rectangles.

Now, for a very small rectangle, whose outline fits neatly within the field of view, the recognition process is a bit different. Such a rectangle is recognized by the excitations of primitive components – corners and lines of specific orientations – at relative locations on the retina. The next-higher level of the visual cortex recognizes the simultaneous occurrence of all four corner orientations along with horizontal and vertical lines. It is still an invariant representation, because all small rectangles will produce similar collections of corners and lines, even though they may be recognized at different locations on the retina.

The neural sites corresponding to all the various respective instances of recognition regarding large and small rectangles become linked at a higher level as a more generic and abstract representation of a rectangle. How do they become linked? Well, think of walking toward some rectangular object, say, a television or a doorway. The object, when it is far away, fits easily within the field of view, and hence, is 'small'. As the object gets closer, it gets 'bigger' in the field of view. While progressing from small to big, there will come a point at which the detection of small rectangles overlaps with the detection of big rectangles. At that point, both sorts of detectors are simultaneously active, and thereby become linked at a higher level. Further, while the rectangle grows in the field of view, or moves around, the various neural sites corresponding to similar but different sizes and locations of rectangles are simultaneously active, and thereby, become linked at a higher level. That higher-level node is then available as a representation of all rectangles of a particular aspect ratio, big and small.

It seems that invariance is a natural virtue of the hierarchical structuring of layered abstractions. Whenever the output of one cortical area forms the input to another cortical area, another order of abstraction occurs. The human cortex automatically organizes level upon level of abstractions in the recognition of very complex patterns of patterns of patterns. The concept of a 'black hole', for example, is extremely abstract. It is built out of many slightly more concrete concepts, of which they themselves are built out of concepts that are yet even more concrete. Every concept, no matter how abstract, is ultimately resolvable down to relationships between real-world patterns of reliable and repeatable empirical evidence.

Exclusive Orthogonalization

The dimensions of nature are relatively independent of each other. Under typical circumstances, there are no overlaps or projections between any of them, including the three dimensions of space, the dimension of time, the dimension of mass or the dimension of energy. For instance, we can move in the X direction of space without moving at all in the Y or Z direction. The dimensions of space are said to be *orthogonal*, which implies that there is no translation – no sort of equivalence or conversion factor – from one dimension to any other (relativistic effects notwithstanding). I want to make the case that the brain tries to partition its conceptual space into orthogonal dimensions.

Just as we earlier saw that a group of people can make better decisions if its constituent minds are diverse and independent, we'll now see that the same is true for groups of pattern detectors within a single brain. A brain's level of intelligence depends on its constituent detectors being diverse and independent – orthogonal.

Recall the similarity between conceptual hierarchies and the Fourier Transform. It turns out that there is an interesting lesson we can learn from the theory behind the mathematical formula – a lesson that applies to the organization of the brain. The theory behind the Fourier Transform demands that the basis vectors – the primitive elements of pattern that are to be picked out of the input signal – must be mathematically orthogonal. And, indeed, sine waves of different frequencies are. In a similar manner, the brain should try to organize all the various low-level primitive patterns it uses to mentally describe the world so that they are as orthogonal as possible. Our goal in this section is to understand how it does that.

We have already seen how neurons can organize themselves to become sensitive to a pattern that occurs frequently. But, we certainly don't want many neurons at a particular level all to be sensitive to the very same pattern. We want each neuron to be sensitive to just one particular pattern. We want all the neurons at any particular level to differentially recognize a fairly large library of orthogonal patterns. So, how can we get neurons to automatically divvy themselves up into different classifications of patterned sensitivity? How can we get them to be orthogonal? The answer relies on the nonlinear process of lateral inhibition, which has the added effect of balancing neural activity on the edge of chaos.

Recall that some neurons are excitatory and others are inhibitory. Now, consider what happens in a small patch of cortex when a newborn baby perceives a new pattern for the first time. Among the many neurons that are eligible to respond, there are a few lucky ones scattered randomly throughout the pattern that happen to respond before the rest. It is just dumb luck that they happen to be slightly more attuned to the pattern than their neighbors. So they fire first. But at the same time, suppose their firings cause inhibitory signals to be sent to all their neighbors. The inhibitory signals convey the message, in essence, "don't worry, I've got this pattern; I'm already attuned to it more than the rest of you are."

When a different pattern lands on the same region, a new and different set of sparsely distributed neurons comes to life, immediately locking out neighboring neurons. A single neuron may initially participate in the recognition of multiple patterns, but as time goes by, it will settle into some particular pattern, becoming more and more specific in its tuning. A neuron will eventually resolve any ambiguities in its detected patterns by tightening its allegiance with one pattern or another. And by that mechanism, various levels of cortex will automatically divvy up their classifications into the most primitive and orthogonal patterns possible.

Let us consider the mechanism in slightly more detail. Whenever an excitatory neuron fires, it causes many inhibitory neurons, whose dendrites touch it, to fire also. The muting effect from those inhibitory neurons prevents any other excitatory neurons in the local region from firing. They act in the capacity of an automatic gain control, ensuring that at least one neuron fires, but not more than a few, within the local region. The size of such a region is defined by the typical expanse over which inhibitory dendrites will stretch. Let us refer to this proximal area, which surrounds each and every excitatory neuron, as a *region of orthogonality*. The number of different pattern types recognizable within a typical region of orthogonality is related to the number of neurons within its coverage area. The mechanism of lateral inhibition ensures that there is a best-guess made within a particular region of cortex for every pattern to which it is ever exposed.

For a particular region of orthogonality and a particular pattern of input, the neuron that happens to be the most sensitive will fire first and thereby cause the inhibition of all the rest of the neurons in that region. At the same time, the firing of that neuron will further strengthen all its connections with the particular photoreceptors that caused it to fire. By firing, the neuron has declared itself as responsible for recognizing the incoming pattern and at the same time has become more sensitive to that pattern. This is referred to as a 'winner-take-all' strategy.

Given this winner-take-all strategy, what sorts of patterns should we expect to see at the lowest level of the visual cortex where neurons are one step away from photoreceptors? We can gain some insight by conducting an experiment. Punch a small hole in the center of a piece of paper. Now, look at the world through the hole as you hold the paper about a foot away and move it around. This simulates the sort of visual fields to which the dendrites of a first-level visual cortical neuron are exposed during typical experiences of vision. What you'll see, if you actually do the experiment, is mostly solid or shaded colors, with the occasional occurrence of an edge moving through the small field. So, for a region of orthogonality in the first-level visual cortex, there will be some neurons that claim the duty of representing various solid colors, some others that represent shadings, and still others that represent edges.

As a region is exposed to more and more visual experiences, it will encounter edges of all orientations. A region will likely have many edge-detecting neurons, each sensitive to an edge at a different angle. The same is true for various orientations of shading. Indeed, brain researchers find edge-sensitive neurons when they probe visual areas of the cortex (although, at least some detection of edges may actually take place in the retina).

While surfaces can be plain, they can also have various textures. So in addition to the sorts of detectors already mentioned, there are likely many sorts of texture pattern detectors. And while edges are often straight, they can also be curved or can end at a corner. So, in addition to straight edge detectors there also may be curved edge detectors and corner detectors. However, depending on various parameters, the detection of curves and corners might take place at the next-higher level by the detection of edges of particular orientations arranged in particular patterns. The same must be true for gross characteristics of texture as well.

To roughly simulate the experience of a neuron in the second level of the visual cortical hierarchy, punch a cluster of holes in the center of a piece of paper and look through the cluster as you move the paper around. Even though each individual hole is usually a solid color, or an occasional edge, the cluster of holes begins to reveal the higher level concepts of corners and curves. The neurons in the second level need only look for familiar patterns built from clusters of primitive patterns defined at the first level. Of course, unlike our trivial experiment, there is significant overlap between the clusters of regions in the neural architecture that ensure orthogonality within them. And so, the automatic partitioning of familiar patterns within many overlapping regions of orthogonality results in similar detectors being randomly 'sprinkled' across the entire visual field for each and every primitive pattern.

As I previously mentioned, we can't really know what actually constitutes the entire library of orthogonal visual primitives until we build an orthogonally-partitioning, coincidence-detecting visual system and expose it to real scenery. Until then, it will be even more difficult to speculate on what the second-order and higher-order relationships actually are. But, all we have to do is build it, feed it lots of image data, and allow it to automatically partition itself. There has been, in fact, a lot of research going on in similar styles of architecture, mostly directed toward efficient image compression.

Foundations of Mimicry

We now have a cognitive mechanism that holds great promise for supporting generalized mimicry. It emerges naturally from a neuron's ability to mimic or represent some aspect of the environment, through simple Hebbian learning. When properly organized, a brain full of neurons achieves a synergy that allows it to represent many aspects of the environment in terms of abstract patterns. The very abstract nature of those neural representations is what allows our minds to mimic the behaviors of others.

Just as the next note of a song automatically pops into the mind of one who is singing it, so do various patterns of behavior automatically occur to us under the right circumstances. And, just as the relative intervals between the notes of a song can be sung in any key, so can memetic behaviors abstractly apply in many different but similar situations and to various people.

When a young boy watches his father perform an often-repeated behavioral sequence, the boy's brain builds a prediction chain corresponding to that sequence. If the father tends to perform the behavior only under certain circumstances, then the behavior will be linked to the recognition of those circumstances. Whenever the boy finds himself in a similar situation, the behavior will automatically occur to him, simply as a result of his recognizing the same circumstantial patterns. Since the behavioral actions in the sequence are abstractly encoded, they can apply equally well to performance by him as by his father. And, they can involve similar tools and objects in his environment as were in his father's environment. Such is the beauty of abstract invariance.

If a boy sees his father tuck a napkin under his belt every time he sits down in a restaurant, the boy's brain remembers the behavior in connection with restaurants. Whenever the boy finds himself sitting at a table in a restaurant, the thought will automatically occur to him to tuck his napkin under his belt. It doesn't matter what restaurant he is in, what color the napkin is, or what belt he is wearing; he automatically recognizes the functional similarity between them all on the basis of how they each map to identical or proximal nodes in the cortex through linkages established by experience, and he acts accordingly.

We can now begin to see how a child might learn to react to its environment in a way that mimics its parents' reactions to similar environments. But we really need to understand how a human mind can be both proactive and creative. The answer lies in the mind's ability to simulate the environment within its imagination, and in its propensity to contemplate (automatically simulate) many different scenarios within that imagined environment.

If a contemplated behavioral sequence requires significant up-front sacrifice, a human mind might first imagine the performance of such a behavior to see if it is likely to yield the expected benefit. If the imagination of the sequence does not produce the desired outcome, then the mind might iteratively imagine various mutations of the sequence in order to find one that *is* expected to yield benefit. This human ability of imagining the future outcomes of various actions is the nature of proactive and creative intelligence. It is a matter of repetitive simulation, mutation, and evaluation – an evolutionary process – which is the focus of the next chapter.

Chapter 8. The Creative Mind

By characterizing the fundamental elements of intelligence completely in terms of patterns – spatial and temporal relationships – we have enabled a memetic perspective on intelligence that is much clearer than any other. From such a perspective, we see that even the simple act of mimicry can be intelligent. Indeed, by mimicking the various behaviors of a very intelligent person, under circumstances identical to when that intelligent person performed them, the mimicked behaviors may be considered to be just as intelligent as the authentic ones. All surgeons, for example, learn their skills by watching and mimicking experienced professionals. We certainly consider surgeons to be intelligent, even though their actions result mostly from simple mimicry. But, while mimicry may sometimes be considered intelligent, it is not typically creative.

Now that we have discovered how memetic relationships are encoded by connection patterns in the brain, we are finally prepared to describe a model of human creativity that is defined completely in terms of memes. The key to understanding creative intelligence, same as the key to understanding all aspects of life, is, of course, the prerequisite understanding of evolution. Indeed, we are finding plenty of reasons to suspect that the prime mover of all intelligence is, fundamentally, a process of evolution. Human intelligence involves a slightly unique style of evolution characterized by its Lamarckian nature, but it is a process of replication, mutation and selection, nonetheless.

When patterns are replicated, mutated and differentially selected, they creatively evolve toward new and better patterns. Such is the process by which biological evolution achieves its creativity, and also by which deterministic human minds proactively achieve their creativity. But, before we attempt to identify the creative evolutionary processes operating on replicated neural firing patterns in human brains, let us first be absolutely clear as to what we mean when we speak of human creativity.

Creativity

Our vivid conscious experiences encourage us to model human creativity as something that can never be automated. We sit and we think about things, and then, ideas pop into our thoughts. It certainly feels like magic. But any attempt to elevate the human mind above deterministic physics is tantamount to introducing a spirit. So, let us try to analyze exactly what it is that we are trying to automate. Let us consider some clear cases of human creativity. What we'll find is a process best summarized as a heuristically directed search for a string of complementary causal memes spanning the gap between what exists and what is desired.

Let us contemplate how creative advances in technology might occur through an evolutionary process operating on strings of causal memes. The model we'll use for memetic evolution is similar to the model we described for genetic evolution. Memes, like genes, move about through their environment until they meet-up with complementary partners. Their complementary interfaces allow them to join-up and form new co-adapted entities. Complementary memes can serendipitously meet-up in a brain, causing something of an unexpected epiphany, or they can be sought out through a laborious search of relevant literature. Complementary memes of technology are more likely to meet-up in the minds of scientists simply because it is they who collect technology memes, and it is they who have become good at mentally simulating and assessing the effects of various meme combinations.

Through the trial and evaluation of various causal strings, we automatically evolve our ideas. While such a process of evolution often requires empirical evaluation, it can sometimes happen completely within a virtual environment of mental simulation. Indeed, our human brains seem to be built to simulate the real world in what we commonly refer to as *the mind's eye*, or the imagination. But, before we consider the creativity that can result from evaluation by mental simulation, let us focus on the simpler process involving empirical evaluation.

Consider a classic case of creative genius – the case in which Edison conceived of an idea for an electric light bulb. We may logically speculate that at some point in Edison's life he noticed how some electronic components, such as thin wire filaments, could be caused to glow by passing large amounts of electric current through them. Edison then had a concept of causation – a causal meme. It was nothing more than a particular relationship between an electric current, a filament and the production of light. He undoubtedly had many other concepts of causation, including one for generating an electric current. Such a causal meme for generating electricity would have been highly complementary to a causal meme of passing electricity through a filament.

We may further speculate that on another occasion, while conducting his work at night, Edison might have wished he could somehow have more light than what he was getting from candles and lanterns. His desire for light established a goal to which a complementary causal meme for producing light would nicely attach. His brain effectively used reverse-engineering to get from the desire for light back to the idea of passing a high current through a filament to produce light. Recall that reverse-engineering merely entails a search for any causal meme having an output pattern that matches the desired goal. By identifying such a causal meme, his brain was able to conceive of a 'brilliant' idea.

I have previously analogized causal memes as something like dominos or like jigsaw puzzle pieces. To span a gap in a puzzle with a single row of pieces, we need to find pieces that have complementary sides, allowing them to be snapped together. We need to get from one particular style of side on the left to another particular style of side on the right. Analogously for causal memes, we need to get from the state that exists to a desired goal state. Like puzzle pieces, causal memes have patterned interfaces – inputs and outputs – that need to be matched up in order to be useful. And the patterns of those interfaces can be very abstract.

Edison's creative process started with reverse-engineering, by finding a causal meme with an output state matching the goal state, but there was still a lot of work to be done in the forward direction. He soon discovered that a thin wire filament heated by an electrical current would quickly burn up in the presence of oxygen. He needed to find ways to make the light brighter and to keep the filament from burning out so quickly. He needed to mutate the general relationship, over and over, and evaluate each of those mutations. One of the styles of mutations he used was to consider different types of filaments.

Unfortunately for Edison, he could not use his internal mental simulator to evaluate the various mutated relationships. Since he hadn't had sufficient experience with various types of filaments, he needed to determine the best one by empirical evaluation. So, he exhaustively tried many different types of filaments through a laborious process of evolving iteration, discarding the bad ones, keeping the better ones. Of course, such a style of differential selection is very similar to the process that nature uses to evolve better forms of biological life.

Edison conducted many hundreds of experiments. The process was tedious and at times yielded only incremental improvements, but the diligence paid off. He finally settled on a tungsten filament in an evacuated chamber of glass. The glass chamber kept oxygen away from the hot filament while allowing light to escape. His previous knowledge of various ways to evacuate glass chambers represented another set of causal memes on which he could draw, through reverse-engineering, to backwardly span the gap from his desire to prevent oxidation back to the situation that existed for him. Creativity usually starts by the backward process of reverse-engineering. But optimization generally requires forward evolution.

We like to think of creativity as something that seems to occur like a lightning bolt from the blue - an epiphany. But the truth is quite different. Creativity, like finding that long-missing puzzle piece, is more likely to result from a diligent intellectual search of relevant patterns. And, not surprisingly, creativity is more likely to happen in the minds of educated people simply because they carry around more causal memes. The more memes one has in one's brain, the more meme combinations one can imagine.

There are many people today who are equipped, by virtue of their heredity and environment, to combine memes in a creative manner. And there are plenty of social pressures and enticements for them to do so. The social pressures and enticements prey on their emotions in ways that cause those entrepreneurial types of people to experiment, to invest time and money, and to create businesses that invent things. Those social forces applied to Edison as well. He embarked on the inventive process because of some string of causal relationships in his brain that made him think it was possible and worthwhile.

While Edison had little previous experience with the relevant characteristics of various types of filaments, which therefore required him to empirically evaluate them, there are in fact many domains of human thought in which our mental simulators are perfectly capable of doing all the forward evolutionary iterations. Consider, for example, the game of chess. A chess player might first identify a good move by the strategic process of reverse engineering, working backward from the opponent's king to the possible pieces that might be able to capture it. But to verify that a move is indeed a good one requires an exhaustive search of the many levels of intervening tactical responses, through many forward simulations of both players' possible moves.

A good chess player routinely simulates the effects of various possible moves, all within his own mind. A player will think to himself: "If I move my pawn, my opponent will likely respond by moving his rook or his bishop; I shall then respond to his rook move by moving my knight, or to his bishop move by moving my queen." The various scenarios considered by a good player may go as many as five or ten moves into the game. They represent simulated possibilities of how the game might likely proceed. By imagining the likely outcomes of the various possible moves, each move takes on a different value of expectation. The various moves compete for selection on the basis of those expected values.

Many mutations on a single line of play are able to be simulated within a player's mind, and a selection process keeps track of the line of play that has so far been identified as the best. In that manner, the best move evolves over time within the player's brain, through a simple process of probing, simulating, evaluating and selecting. The game of chess is easily simulated, computationally, and thereby allows a process of evolution to be applied within that simulated environment. Computers have gotten very good at the game of chess – now better than humans – by using this very technique of applying a systematic style of evolution within a simulation of the game.

Whatever we are able to accurately simulate on a computer, we are then able to predict its likely future through that simulation. And, whenever we can predict the future through computational simulation, we can find intelligent solutions just by simulating and evaluating many various plans.

Simulation

All great human accomplishments spring from wild ideas that, in the minds of their conceivers, just might be possible. But, before a wild idea can be deemed a good idea, there has to be a mental simulation that justifies it as potentially yielding future benefit. The principle is valid for all creative thinking, whether it involves wild innovation or just subtle improvement. A human's ability to simulate the world within its thoughts is what enables an evolutionary process to occur within its mind.

When a typical human mind contemplates something, it goes over and over the same thoughts, searching for ways to modify or mutate the best available option in hopes that it may become a better option. And when a mind finally discovers a new option that seems better than all the others, it must have already assessed the new option as being better by submitting it to mental simulation.

The mental simulation of reality has gotten quite difficult in modern times because the world is now a complex place full of very complicated relationships. Consider the world of a businessman who must decide how he will manage his business. He knows that in order to make more profits he needs to sell more product. And to sell more product he needs to advertise more. So, he has two conflicting causal memes. One meme leads him to believe he can raise profits by spending more money on advertising. But another meme tells him that the very spending he does on advertising directly reduces profits. So, it is a complicated relationship.

The same is true for product pricing. The higher the selling price, the higher the profit per item sold. But, raising prices leads to fewer sales, which has the effect of lowering overall profits. Like advertising, proper product pricing depends on a complicated relationship between several different factors. And, further complicating matters, the effectiveness of his advertising heavily depends on the price he advertises. It is very difficult for a businessman to confidently imagine what will happen under various scenarios of advertising and pricing.

To resolve these dilemmas, a businessman might use a spreadsheet in which he can mathematically model all the relationships of his business. He uses the spreadsheet to simulate reality. Then, he can twiddle with various parameters to see how the changes cause overall profits to rise or to fall. He pseudo-randomly mutates the parameters in an effort to maximize profits. Once he finds a combination that yields good results, he sticks to that general theme, but adjusts parameters at a finer level. The twiddling of parameters is directly analogous to the way in which genes explore the biological landscape of life. When genes find a species that is able to survive within some ecological niche, then smaller mutations from meiotic recombinations have the effect of fine-tuning the species.

Spreadsheet models are analogous to the simulators within our brains. Our brains are designed so that various mutations of scenarios are automatically submitted to our mental simulators to see which ones are most likely to yield good results. By such a mechanism, our thoughts evolve.

The big difference between genetic evolution and memetic evolution is that the internal mutation of a meme can be Lamarckian. Given a particular goal, the method for achieving the goal can be reverse-engineered by searching for a known causal meme having an output that matches the goal. Unfortunately, our mental process of reverse-engineering cannot always be counted on to yield the same precision as is yielded by our methods of forward simulation. But, once a close idea is found, then forward simulation can be used, through evolving iteration, to get more precision.

To analogize, consider a common technique for mentally finding the square root of some number, say, 105. A reverse engineering process tells us that the answer is about 10, because we already have a causal meme stating that 10 squared is 100. The output of that causal meme nearly matches our goal. We can then try forward simulation on mutations of 10 by squaring 10.1, 10.2, 10.3, and so on, to get better precision. The reason for the imbalance in precision between our abilities to run simulations forward and backward is because we usually witness the state of the world propagating in the forward direction of time.

In recent centuries, mathematical formulas have memetically emerged to aid our mental processes of simulation. Indeed, most mathematical formulas in the hard sciences are intended to simulate some reliable aspect of nature. And we routinely program our computers to iteratively calculate those formulas, thereby simulating the passing of time for various sorts of virtual systems. Consequently, the real power of simulation, and hence prediction, lies in the realm of computers. As computers continue to gain power in future years, we will increasingly use them to simulate the processes of life. We will use them to simulate everything from molecular interactions to social dynamics, and who knows what else. We have indeed just begun to use computers to simulate the folding of proteins, resulting from the sequences of amino acids in them. This is the first step in the simulation of life.

In the very near future, we will undoubtedly learn many new principles of evolution by enabling simulated life to evolve within completely virtual, simulated universes, whose laws are enforced by the deterministic computation of computers. This is a nearly indisputable fact that is critical for understanding the memetic perspective on cosmology that I'll later present.

Here is an interesting and relevant philosophical question: Is it possible for simulated life to ever achieve consciousness? To answer that question we could go back to the previously developed definition for consciousness, which, at its essence, only requires for a mind to *believe* it feels consciousness. But, you may not agree with those conclusions, and indeed, perhaps I've gotten it wrong. So, let's go to a deeper level. Let us imagine the mathematical simulation of a human brain all the way down to the atomic level. Let us imagine that we can somehow know the entire sub-atomic state of a given mature brain, and that we can load that data into a hugely powerful computer capable of mathematically simulating sub-atomic quantum processes. Then, just press 'Enter', and watch the computer screen as the simulated brain begins to think.

Would such a simulated brain experience consciousness? The answer is very clear in my mind. If it didn't experience consciousness, then the simulator is simply not accurate. If the simulator is accurate, then the simulation must produce consciousness. To conclude otherwise presupposes the simulator's failure to simulate some sort of supernatural stuff, like a spirit.

If recent trends in the development of cheap computing power continue, we may, in the next thirty years, produce computer-simulated life that is self-aware and in fact conscious, at least to the same extent we humans consider ourselves to be conscious. Such conscious life won't know it is simulated unless we feed it data, into its sensory input stream, that allows it to know it is simulated. But why would we ever do that? It would defeat the purpose of simulating life. We will inevitably simulate life solely for the purpose of learning about life. Let's just feed a simulated mind streams of sensory data as if it exists in a world such as ours. Let's allow it to believe it is holding a pencil and a crossword puzzle. Then we can monitor its neural flows and learn how it achieves creative thought. My bet is that we would find one or more evolutionary processes operating at the level of neural excitation patterns.

The Internal Evolution of Ideas

Suppose there is no source of creativity in the universe other than evolution – no omnipotent God, no soul, no volitional free will – just evolution operating at many different levels on many sorts of replicators. From the perspective of Ockham's razor, such an extremely lean theory is more scientifically credible than any other that currently exists. Although very non-intuitive, the parsimony of science demands we reject theories that include a creative soul or a free will. It is much simpler to conclude that evolution accounts for all creativity. And so, it is now time to consider the evolution of ideas internally, within a single mind, as a source for automatic individual human creativity. We start with some loosely related anecdotal evidence in lower forms of life.

Let us return the focus of our thoughts to experiments described by psychologist B. F. Skinner in which animals establish cause-and-effect relationships between actions that they perform and rewards that are given to them in response to those actions. Recall the 'Skinner box' in which a pigeon *learns* how to get more food by pecking at a certain button.

Skinner reports on a variation of the experiment in which he discontinues the relationship between the button and the reward. Instead, Skinner gives the bird a reward on an occasional basis, independent of the bird's actions. At that point in the experiment, according to Skinner, many of the birds build up a 'superstitious' behavior that varies from bird to bird. The birds apparently come to 'believe' that whatever they happened to be doing just before the last reward appeared is what caused the reward to appear. So, a bird might jerk its head to one side in anticipation of another reward, and it might continue to do so until the next reward appears, which then verifies in its mind that it truly is the continued head-jerking that causes the reward.

If a pigeon 'believes' a certain behavior causes food to appear, it will 'proactively' replicate that behavior. Over time, Skinner reports, a pigeon's repeated behavior experiences 'topographic drift', which simply means it changes, slowly. Now then, it is obvious that the brains of pigeons automatically select certain behaviors, over others, for performance. Whatever behavior a pigeon expects is most likely to yield reward is the behavior that the pigeon will automatically select. Thus, we have satisfied all the requirements for memetic evolution. The ongoing replication of a behavior, along with the topographic drift and the selection of some behaviors over others, all point to the possibility of an automatically mutating meme inside the brain of the pigeon.

While the behavior of the bird in the Skinner box is not copied from another bird in the manner of memes discussed so far, the determining patterns of behavior do seem to possess the dynamic characteristics of replication, mutation and selection, completely within the single brain of the pigeon. And, whatever pattern of neural activity causes the pigeon's replicated behavior, it seems to serve as a template for its own replication every time the behavior is repeated.

I certainly don't mean to suggest that pigeons think like humans. Whatever sort of evolutionary process might be going on in the brain of a pigeon, it must be rudimentary compared to that of a human. Nevertheless, we must suspect that Hebbian learning is a characteristic of all brain cells, including those of pigeons. And the very nature of Hebbian learning is to direct neural activity down paths it has already gone before under similar environmental circumstances. By creating and reinforcing neural pathways, neural activity thus serves as a template for its own replication or translation. Now, having suggested the possibility of an evolving inter-brain meme in birds, we don't have to look far to find clear evidence of it in humans.

Consider a young tennis player who practices his stroke in an effort to improve his game. In all sports, players must practice their skills by more than a million repetitions before they can hope to become world-class athletes. During those numerous repetitions, many minor variations will inevitably occur. The brain then imposes a selection process by evaluating the perceived result of each instance of repetition. In the case of tennis, the selection process evaluates the speed and accuracy of each shot and remembers the muscle contraction sequences of the good ones. In this manner, the player slowly refines the neural mechanism responsible for directing his stroke. The corresponding behavior, or meme, evolves as it is repetitively executed within his single brain.

Collections of neurons responsible for directing muscle contraction sequences must have some random components of topographic drift initially built into them so that every stroke is not identically executed. The random component generates muscle contraction sequences within some probability distribution whose width becomes narrower as the action becomes more proficient. So, a beginner has muscle contraction sequences that vary quite a bit from stroke to stroke, but an expert has very little deviation between strokes. A novice tennis player's stroke would never improve without the little deviations between various strokes. This is self-evident, simply because an absence of deviation would imply that the stroke pattern will always remain the same.

Daniel Dennett certainly recognizes the scientific value of supposing evolutionary processes in the human brain. For example, he speculates in *Consciousness Explained* (1991) that early hominid forms of recursive thought processes might have begun when our ancestors first became capable of talking to themselves. The primitive ideas expressed by a hominid's speech production circuits might have excited the auditory circuits in his very own brain, thereby causing him to speak a new, related idea. The process, being repeated again and again in a conversation with himself, could, perhaps, be described as the precursor to a *train of thought*. This initial style of recursive auto-communication, or something similar to it, might have led to a process of evolving ideas within the brain.

While Dennett is flirting with something very interesting here, I think the roots of an evolutionary thought process go much deeper. As I have already mentioned, the fundamental mechanism for pattern replication within the brain is likely to be the mechanism of Hebbian learning. Recall my suggestion that Hebbian learning creates neural pathways that have the effect of guiding future excitations, under similar circumstances, down those very same pathways – a fundamental basis for pattern replication.

We have already seen how hierarchies flowing upward can naturally result from Hebbian learning. And, we have also speculated on how parallel pathways, going downward, might result from the very same mechanism of Hebbian learning applied to neurons pointing in the opposite direction. The two sets of parallel hierarchical pathways, one going up and the other down, then form circular or nearly circular pathways that enable neural firing activity to be repeated over and over again. Such repeated patterns of neural activity form the basis for creative intelligence through the evolution of thought. We'll further develop this line of reasoning in a moment. But, first, let's look at some other interesting evidence of naturally occurring intelligence.

Consistent with the idea that evolution is responsible for all creativity, consider an interesting example of yet another evolutionary process within the human body that exhibits a type of intelligence. In 1972, Gerald M. Edelman, M.D., Ph.D. received a Nobel Prize for his work on the human immune system, in which he discovered that the production of specifically-tuned antibodies involves an evolutionary process.

> Antibodies have special sites that match or bind portions of other molecules, almost the way a cookie cutter matches a cookie of a given shape. What is remarkable is that practically any foreign molecule or antigen injected into the body will elicit the production of a complementary antibody that is essential for subsequent immune defense. ... The basis for molecular recognition of an enormous number of different foreign molecules is somatic variation in the antibody genes of each individual that leads to the production of a vast repertoire of antibodies, each with a different binding site. Exposure of the enormous repertoire of different antibodies to a foreign molecule is followed by the selection and growth of the cells bearing just those antibodies that fit the foreign chemical structure of a given antigen sufficiently well, even a structure that never occurred before in the history of the Earth. Although

the mechanisms and timing of selective events obviously differ in evolution and immunity, the principles are the same – the Darwinian process of variation and selection.

   – **Edelman and Tononi**, *A Universe of Consciousness* (2000, pp.82,83)

Through the random production of millions of lymphocytes, many styles of antibodies are formed. The various antibodies compete in a process of selection based on the degree to which they are complementary with the antigen. The antibodies that are best able to latch onto the target antigen are recognized by the immune system as successful and are subsequently reproduced. Through this adaptive and creative mechanism, the human immune system is capable of intelligently generating appropriate antibodies to fight brand new, never-before-experienced diseases. If we are looking for evidence to support the idea that evolution accounts for all intelligent creativity in the universe, the human immune system is another good example.

Years after he proposed his Darwinian vision of immunology, Edelman followed up with a Darwinian theory of self-organizing neural systems. In a book called *Neural Darwinism* (1987), Edelman showed how neural connections evolve through competitive selection. For a given sensory input pattern, many neurons compete for the honor of representing that pattern. Among those many neurons, 'fit' neurons become fitter, and unfit neurons are inhibited or pruned. Edelman refers to all the winners, for a given pattern, as a neuronal group. Neuronal groups that recognize common patterns tend to grow stronger, more sensitive and more discriminating. Thus, as Edelman points out, the self-organizing principles of Hebbian learning are more properly viewed within the context of an evolutionary process.

At all cortical levels, connection patterns that happen to already reflect the coincident relationships expressed by incoming sensory patterns win by becoming further strengthened. Survival of the fittest is, of course, a common evolutionary theme. As a simple example, a popular path worn through a forest gets even wider and more worn from its continued use. The wider it gets, the more popular it becomes; and the more popular it becomes, the wider it gets. Even among the competing businesses in a free-market system of commerce, the ones that best match the needs of society gain more profits and get stronger as a result. The rich get richer and the poor get poorer – the fit get fitter and the unfit go extinct. It is true of genetic lineages, true of businesses, true of paths in a forest, and true of neural connections.

With regard to Edelman's theory of Neural Darwinism, it seems the plodding biological evolution of genes has found a way to create a more efficient style of evolutionary process, operating at a higher level and proceeding at a much faster pace. That second evolutionary process automatically connects neurons so that they become a simulator of the sensed environment. But we're not done yet. In addition to the biological evolution that designed the human brain, and in addition to Edelman's evolution of neural connections, there is yet another Darwinian process on top of them both – a third evolutionary process – operating within the mentally simulated environment that the lower Darwinian processes create.

In this highest of evolutionary processes (so far), iterated plans are differentially imagined (simulated) and thereby evaluated in terms of the various amounts of pleasure or happiness they are expected to yield. Plans are differentially selected on the statistical basis of how well they are likely to perpetuate the genes that enabled the lowest-level Darwinian process of biological evolution. Let us focus our attention on this extremely fast, high-level style of cognitive evolution from which our evolving thoughts emanate.

We need to get used to the idea that there are at least several somewhat different styles of evolution. For example, unlike the evolution of DNA, the evolution of neural activity replicates in time, not space. Whereas replicated genes are physical entities that persist in multiple copies of matter for a long while, replicated firing patterns are very transient in nature. They last only long enough to be evaluated. Keeping in mind that a firing pattern can represent a plan of action, once a firing pattern is evaluated, the amount of happiness it is expected to yield is compared with that of the previously favored plan. If the new firing pattern predicts a greater amount of expected happiness, it then replaces the previously favored plan and becomes the new favored plan. So, only the favorite plan is allowed to persist as something of a resonating firing pattern. All other firing patterns quickly dissipate their energies as the train of thought shifts from one idea to the next.

This is a serial style of evolution, as opposed to the parallel style of evolution exhibited by the many species of biology. A serial style of evolution is like dropping a single mouse into a maze. It randomly explores the many corridors, one at a time, and, given enough time, it eventually finds the cheese. A parallel style of evolution is like dropping many mice into a maze. Just as many biological species conduct a parallel exploration of ecological niches, so can many mice explore many corridors simultaneously. And they are likely to find the cheese much more quickly than could a single mouse.

Even though human cognition uses a serial style of evolution for creativity, it employs an extremely fast implementation of that process, in comparison with the slow implementation of the parallel evolution of biology. Computers likewise tend to use a serial style of evolution in their methods for achieving artificial intelligence, but their transistor-based implementations are even far faster than the human, neuron-based implementations of cognition.

Repetitious patterns of neural firings, going up and down various connected hierarchies of Hawkin's memory-prediction model, give us just the sort of pattern replication we need to support the higher-level style of cognitive evolution. But, remember, it is replication of patterns in time, not space.

Now that we have a plausible mechanism for replicating patterns of neural flow within the cortex, we need a mechanism for mutating those patterns. Such a mechanism had better lead to a Lamarckian style of mutation. That is, it had better produce mutations that we would consider to be more intelligently produced than what would be expected from random mutation.

What we are looking for is an ongoing process of 'trying out' various combinations of complementary memes so as to find a string that spans a gap of causality in between the current state and some goal state. Once a string of memes is found, it can be optimized by systematically replacing some of its causal memes with others having similar causal effects. This sort of process is reminiscent of the meiotic shuffling of genes accomplished by occasionally replacing them with their respective similar genetic alleles.

Consider an example. If I need money, there are two broad styles of causality that can satisfy the goal: I can earn it or steal it. Both plans have similar inputs and similar outputs (ignoring for now the possibility of getting caught when stealing). But, while the plans have very similar causal effects, they have vastly different intervening behaviors. Further, each of these broad categories holds a variety of possible plans within them. For instance, if I am going to steal, I could rob a bank, 'knock over' a liquor store, burgle a house, or kidnap the child of a wealthy family. If, on the other hand, I insist on earning the money, there are hundreds of different job listings in the newspaper. All these various plans represent memetic alleles for the causal act of getting money.

All the plans are similar in their causal effects: They all start with an input condition of having little or no money, and they all produce an output condition of having plenty of money. But they all achieve their similar causal effects through vastly different sequences of behaviors. Thus, the input and output conditions of all the plans map to the same respective hierarchies in conceptual space, but their causal pathways through that space, from input to output, differ dramatically. We want to conceive of a neural architecture, and corresponding evolutionary process, that automatically considers all the various pathways. The question is: How do our plans get mutated by swapping in and out various similar causal components?

Neurons essentially mutate the signals they process in several ways. One of those ways just happens to produce the effects we are looking for. The critical mechanism of conceptual mutation results directly from a neuron's inability to sustain its maximum firing rate for long periods of time. This 'fatiguing' effect guarantees that loops of neural activity will attenuate over time, thereby allowing one idea to give-way to another similar idea in a constantly evolving train of thought. Now that we have a candidate mechanism for flipping from one meme combination to another, let us consider the automatic means for intelligently choosing which of the various combinations of memes to 'try out'.

The act of combining complementary memes is the basis for creating mutated variety among patterns of ideas. It is something of a meiotic process in its ability to combine large, functionally-significant components at the conceptual level. We will soon discover just how the process of combining memes into bigger memes is intelligently directed by the natural organization of conceptual hierarchies in the brain. And, we'll further discover just how that style of mutation is Lamarckian in its ability to reverse-engineer a problem.

Searching through Meme Combinations

Memes can combine with other memes in various ways to express all sorts of complex concepts and behaviors in much the same way that genes combine with other genes to express various sorts of biological matter. Just as the biological process of meiosis 'tries out' various combinations of existing genes, so does the neural architecture of the cortex 'try out' various combinations of learned causal memes by imagining their combined causal effects. For both sources of creativity – biological evolution and human thinking – the underlying mechanism involves a repetitive process of generating a variety of trials, which are then evaluated.

From the perspective of an individual who is engaged in deep thought about a particular problem, the conscious mind clearly repeats its thoughts over and over again in its search for a solution. Speaking from personal experience, I myself sometimes get stuck on a problem, as if my thoughts continually retrace the same 'groove' in the conceptual landscape. Let us speculate on what is happening at the neural level, during those moments of recursive thought. We are specifically concerned with how the neural model presented so far might be capable of identifying various logical options and repetitively probing among them.

As I think about some high-level concept, the corresponding hierarchy in my brain becomes active with firings resonating up and down. The round-trip activity persists for as long as the concept remains in short-term memory. Recall that we have already defined short-term memory to include whatever neural circuits happen to be resonating with circular activity. Indeed, our model presumes that all active and resonating brain circuits determine the totality of the conscious experience. It is what Edelman refers to as the "dynamic core."

Perhaps the circular activity of resonating brain circuits is related to the repetitive process of creating trials. Indeed, I'll endeavor to show that looping neural activity forms the basis for an automatic mechanism of 'trying out' various combinations of memes. That neural process corresponds to the conscious experience of conceiving various new plans and ideas. Let us speculate on how such an automatic process might work.

The automatic cognitive process of combining and recombining memes depends on the crucial ability to quickly identify all memes that are conceptually similar to a given meme. That ability enables memes that are similar to be swapped with each other, in and out of the 'dynamic core', similar to how genetic alleles are swapped in and out of chromosomes. To simplify the task of explaining how this happens, we'll initially consider the simplest type of meme, consisting of only a single hierarchy. Once we understand the basic philosophy, it can be easily extended to causal memes, each of which consists of at least two causally linked hierarchies – one for detecting the input state and one for describing the result of the causality, the output state.

Let us return to the example of Edison's light bulb and the causal effect of passing an electric current through a filament to produce light. The filament, considered by itself, has no associated causality and therefore can be represented by a single hierarchy that merely links together all its measurable characteristics into an abstract set of nested relationships. With regard to Edison's initial idea for producing light, we may presume that his mind automatically conceived of many variations on the original theme simply by swapping between many conceptual hierarchies representing things similar to the original filament.

Now, here is the critical point in explaining how the brain conceives of new ideas: Similar conceptual elements, such as similar sorts of filaments, are guaranteed to have similarly configured hierarchies, and, as such, will tend to share many nodes and neural fibers among their respective branches. As thinking comes to dwell on one of several similar concepts represented by one of several similar hierarchies, neural resonance activity excites that hierarchy entirely, but at the same time, excites all the other similar hierarchies partially. Those similar hierarchies become primed for action by virtue of the activity in the nodes they share with the currently active hierarchy. Eventually, as the active hierarchy experiences the effects of fatigue at higher levels, there can be enough activity in the shared fibers to cause the conscious focus to shift from the current concept to a similar concept – from the current hierarchy to a similar hierarchy. It is the upper-level nodes that shift, and the shifting happens when a heavily firing node fatigues to the point where it can no longer cause the inhibition of other nearby nodes at the same level.

Let us suppose that nodes involved in resonating circuits are able to defy the effects of fatigue if they are resonating with nodes at both higher and lower levels. Since the highest level node of a resonating hierarchy receives firing support from only lower-level nodes, it will naturally succumb more quickly to fatigue than the lower-level nodes in the same circuit. We'll later see more evidence and significance of this effect.

We now have a plausible explanation for how intelligently directed mutation of memes might occur. For instance, if Edison's initial filament were made of iron, his brain would have been much more likely to imagine the trying of tungsten, than, say, a cat's whisker. Why? Because a tungsten filament is more similar to an iron filament than a cat's whisker is to an iron filament. The critical characteristic is, of course, electrical conductivity. The neural hierarchies corresponding to iron and tungsten filaments are likely to overlap more heavily than would the hierarchies for an iron filament and a cat's whisker. The overlapping hierarchies for tungsten and iron filaments will share many nodes and branches having to do with the concept of conductivity.

We will see quite clearly in a later discussion of *induction* that conceptual patterns are deemed similar if they share a lot of neural components in their respective hierarchies. Classes of things are naturally formed by such a metric of similarity. Indeed, the ability to represent similarity between conceptual patterns by the number of nodes they share seems to be the primary functional value of hierarchical representation, even above the efficiency of data compression.

The sharing of neurons and fibers between two *similar* conceptual hierarchies is what allows human cognition to create better-than-random mutations on concepts. It essentially guarantees that two causal memes having complementary interfaces will have their interfaces mapped to heavily overlapping hierarchies, and will thereby become connected, even as the causal concepts are learned. Such consistent mapping from learned concepts into hierarchy space enables the automatic stringing together of many causal elements into complicated yet coherent plans. Many possible plans already exist as intersecting strings in a huge web of interconnected causal elements. But, for a given situation, the best of all the alternative plans can only be identified by activating and evaluating them all, one at a time. The best plan is the one that causes the most activity in whatever genetically defined nodes happen to represent expected future happiness.

Those 'happiness' nodes are likely to directly involve neurons that become active during sex, during the eating of tasty foods, during the perception of healthy and loving children, or whatever other circumstances tend to benefit the genes that define those activities as pleasurable. Or, they may have been learned as 'indirect' happiness nodes by virtue of the causality that tends to lead from them to 'direct' happiness nodes. A complete comprehension of this point relies heavily on my previous discussion of indirect happiness.

Our neural model relies on the ability for a good plan to be remembered during the consideration of alternatives. So, let us suppose that a strong reinforcement from a high-level expectation of future pleasure – characteristic of a good plan – can overcome neural fatigue, and thereby keep the associated plan actively resonating in the background of consciousness. By such a mechanism, we would be able to keep in mind the best plan so far, while continuing to search for a better plan.

There are a couple of other concepts that need to be mentioned in order to put flesh on the skeleton of our neural model. Indeed, we need to refine somewhat our conceptualization of causal memes. I have previously analogized causal memes as dominos that are laid down so that their adjoining ends have matching dot patterns. That is, of course, an overly simplistic model. So, let us extend the model to include 'wide' dominos with the ability to have multiple input patterns and/or multiple output patterns.

The need for multiple outputs becomes apparent when we admit that a causal meme can produce various sorts of potential side effects. For instance, while robbing a bank is a means for getting lots of money, it also yields the possibility of going to jail. The cognitive consideration of such a plan (after it has been swapped into the 'dynamic core') will cause neural activity to occur in whatever hierarchies represent the unpleasant concept of incarceration. Such a possibility for future displeasure makes the plan unattractive, and the fatiguing effects of involved neurons will eventually 'give way' to the activation of a different plan that involves similar causation with respect to getting money, but without the possibility of going to jail.

Now, consider a causal meme having multiple inputs. Recall the discussion of how two hierarchies, one describing a key and another describing a lock, must be active in order to satisfy the complete input for the concept of unlocking a door. Without the presence of a key, the fatiguing effects of neurons involved in the concept of unlocking a door will 'give way' to some other causal concept for gaining entry into a house, such as breaking a window. Indeed, both concepts – unlocking a door, and breaking a window – are similar in the respect that they are likely to share some nodes representing the abstract concept of entering a house.

Realize that the nested relationships of hierarchies are already capable of accounting for multiple inputs and multiple outputs with respect to causal memes. For example, a two-input meme merely implies a single input hierarchy having two distinctly diverse sets of branches collecting data from two wildly different conceptual areas. I simply want to highlight just how the nested relationships of hierarchies can involve many sorts of very diverse concepts. In fact, the two parts of a causal meme, consisting of an input hierarchy and an output hierarchy, are actually just parts of one hierarchy, if indeed they are linked at a higher level. Hierarchies seem to have ultimate power in expressing everything about the universe.

We now have sufficient background to facilitate a discussion on the Lamarckian characteristic of memes. Recall that our proposed framework for neural architecture relies heavily on downward flow of neural activity running in parallel but opposite direction to upward flow. This mechanism essentially makes all links bi-directional. Such bi-directionality provides a basis for the Lamarckian effect of reverse-engineering.

For instance, Edison's desire for light represents a concept that would have mapped to the same hierarchy in his brain as the output of any causal meme for producing light. He did have such a causal meme regarding the passing of electric current through a filament. So, by merely considering the desire for producing light, the downward flow of neural activity could have excited the output sub-hierarchy of the relevant causal meme. By activating the upward-flowing mirror to the output hierarchy of a causal meme, the higher-level representation of causality will also become active. That higher-level node then sends activity back downward to excite its entire hierarchy. The causal concept was suddenly in his conscious thoughts. In this manner, plans can be roughly formulated from the goal, backward. And, once a rough plan is discovered through reverse-engineering, it can be refined through forward evolution.

Neural signals flow up and down conjoined hierarchies searching various combinations of activities. It is an evolutionary process of finding plans that are likely to yield benefit. The process tries and evaluates many combinations of known activities, assessing each for its ultimate causal effect, dwelling on any that promise great future reward in the form of pleasure. In addition to the evolving process of planning, there are other forms of evolutionary processes going on at this level, as well. Indeed, the mere recognition and classification of patterns on the basis of abstract similarities is likely to be an evolutionary process. Let us consider how such a process evolves. Unfortunately, the analysis is necessarily a bit tedious for several paragraphs.

Suppose that a line of a particular orientation exists at some given place in the visual field. As the line is projected on the retina, it excites an area of photoreceptors. In the first level of the cortex, there are thousands of neurons monitoring that small area. Each of those monitoring neurons has dendritic coverage over the same area. That is, each can sense the pattern in that area. Any of those neurons is capable of firing if it is even *close* to matching the pattern being sensed, and since there must be detectors for lines of many different orientations, we may presume that some of those will be close to matching the particular pattern being sensed. So, many similar detectors will want to fire in response to the line being sensed, but most will be inhibited by the previous firing of a more accurately-tuned neuron. So, out of the many neurons that *could* respond to a particular line, only a few that are most accurately tuned to its particular orientation *actually do* respond. Their firings simultaneously inhibit all other neurons that could fire. The fact that there are always other neurons *close* to firing turns out to be very important to the way in which firing patterns evolve within the brain.

A precisely drawn 'I' will always excite the line detectors that are tuned to precisely vertical lines. But humans are capable of recognizing a sloppily drawn '*I* ' as well, even though its primary stroke may not be precisely vertical. So long as it is close to vertical, we want to be able to recognize it. So, for a sloppily drawn '*I* ', the precisely vertical line detectors will be initially inhibited by detectors that are more accurately tuned to the slanted, nearly vertical stroke that actually exists. Keep in mind that the precisely vertical detectors are close enough to fire, and would do so, if not inhibited by the previous firing of more accurate detectors. Now, here comes the important point.

Let us assume that the higher-level recognition of an 'I' requires the perception of a precisely vertical stroke. So, the slanted detectors that initially fire, in accurate response to a slanted '*I* ', will not get any reinforcement from recognition at higher levels. Without reinforcement from above, firing neurons experience fatigue and thereby stop inhibiting whatever closely matching neurons are also capable of firing. So, without any support by downward flow from above, the initially firing, slightly slanted detectors will soon fatigue and give-way to firing by the precisely vertical line detectors that then allow the recognition at a higher level. That recognition will send downward reinforcing support for the vertical line detectors as being 'the best fit' for the familiar pattern of an 'I'. Resonating circuits will then symbolize it in short-term memory as if it were a perfectly drawn 'I'.

I am only telling a small part of a rather complicated story here. The level of tedium prevents me from elaborating any further. I simply want to demonstrate how all the various neurons that are close to firing, for any given pattern, represent multiple channels by which the information of *similar* patterns can be propagated to higher cortical levels in the process of trying to recognize *familiar* patterns. Signals flow up and down, looping over and over again, taking slightly different neural fibers every time, searching for combinations of familiar patterns that match the incoming sensory patterns. It is an evolutionary process of finding the best-fitting familiar patterns for whatever sensory data happens to exist. Such a process is what allows us to recognize the characters of a brand new font, or characters that have been sloppily drawn by hand.

Edelman refers to the existence of multiple neuronal channels, for any representational branch in a network, as *degeneracy*. The redundancy inherent in degenerate systems certainly makes them more robust, but the real benefit to the brain comes from the fact that multiple channels are never exactly the same. The differences between them allow for circular signals to explore nearly circular pathways. Neural flow becomes more 'spiral' than circular.

Edelman uses the term *re-entry* to describe the nearly repetitious nature of signal flow within neural systems. He resists using the word *feedback*, as it implies a non-mutating, strictly circular flow. Circularity implies feedback to the exact point of origin, and as such, is not very interesting. A strictly circular neural flow would cause a brain to think the same thoughts over and over again, forever. But we humans don't quite fit that model of cognition. Instead, as we ponder a difficult problem, we do mull things over and over, but each iteration considers a slightly different viewpoint, perspective, or combination of ideas: What if this, what if that? A more 'spiral' sort of flow implies an evolving thought process that considers many variations on a theme.

Feedback, or re-entry, provides the means for continuously replicating thoughts. It results from the downward flow in Hawkins' model. The predictive expectations that flow down are then able to excite other, similar hierarchical paths that flow back up again, allowing the mind to engage in ongoing speculation of 'what-ifs'. Edelman's notion of *degeneracy* aids the mechanism for intelligently mutating ideas as we mull them over. Each round of rumination excites a slightly different path, or set of paths, in the degenerate set. Given that the paths of a degenerate set are likely to be shared between hierarchies that are very similar in some respect, the resulting mutations are likely to excite similar hierarchies representing memetic alleles. The mutations allow us to 'try out' different combinations of ideas, or memes, completely within our thoughts.

By combining Edelman's concepts of re-entry and degeneracy with Hawkins' model of memory-prediction, we have described a relatively complete neural architecture appearing to have been designed specifically to facilitate the *evolution* of thought. Since this section is perhaps the most important in the whole book, allow me to quickly summarize the important points.

Strings of causal memes combine to form our plans and our recipes. And those plans and recipes are what allow us to manufacture all the wonderful artifacts of technology that enhance our lives. Our brains automatically mutate those strings of causal memes by occasionally substituting memes of similar causality. Any memes having respectively similar input and output interfaces will automatically be candidates for substitution in place of each other. And similarity between two interfaces can be easily identified by the degree to which their respective hierarchies overlap.

Two causal memes fit together, and are automatically connected, when the output hierarchy of one causal meme is sufficiently similar to the input hierarchy of another, or when the outputs of multiple hierarchies combine in a parallel manner to satisfy all the input requirements of another. The complementary memes can then become linked by a concept at a higher level. Their respective hierarchies combine to form a bigger hierarchy. The smaller hierarchical relationships still exist, but there is now a new relationship, represented at a higher level, that recognizes the synergy achievable by combining the smaller elements of causality.

When we are doing a jigsaw puzzle and we find pieces that fit together, we then keep those pieces together, treating them as a new, larger, single piece. In a similar manner, when we discover a group of memes that happen to cooperate well together, we tend to treat the group as a single meme. For example, many disparate memes have combined to create the car, but we can conveniently think of a car as a single type of object. We can even talk about the manufacture of a car as if it were a single act. The same thing happens when we combine many cause-and-effect relationships into a single plan. We then establish in our minds a shortcut representation corresponding to a single cause-and-effect relationship that actually may express many connected causal relationships. Indeed, we can think of going to the store without even considering all the intermediate steps of putting on clothes, locking the house, getting in the car, starting the engine, negotiating traffic, stepping on the gas pedal and so on.

Important technological discoveries usually result from a combination of memes that happen to be complementary. We then conceptually treat such a combination as a new, larger, single meme. We'll soon discuss how the memes of a society propagate from brain to brain through automatic mechanisms, and how they automatically combine and re-combine in many various ways, once they are in those brains, to form new and bigger memes.

The final component for the facilitation of evolving thought is a means for selection – a way of identifying good ideas and rejecting bad ones. The terms good and bad are equated in the brain, respectively, with expectations of pleasure and pain. They represent our *values*. As ideas replicate and mutate, their expected future effects are calculated and predicted, specifically as to whether they will produce pleasure or pain. Those ideas that are expected to yield the greatest long-term pleasure are deemed good, and are selected. Through an iterative process of considering many variations on a theme, the brain finally settles on the one combination of concepts that it predicts will maximize long-term happiness.

Whether or not we have properly identified the precise neural architecture of the human mind, we have at least identified an architecture that is mostly consistent with empirical evidence and seems to be very powerful in its Lamarckian style of intelligently evolving creativity. The bar has been set extremely high for competing theories on neural architecture. I don't see how anyone can deny that some sort of evolutionary process underlies every source of creativity in the entire universe. If you are still not convinced of evolution's ability to account for all intelligence, perhaps one more piece of evidence will make you a believer. So, I shall now discuss how ideas naturally evolve at the cultural level, among many cooperating minds.

The Ultra-Creative Societal Mind

I have already mentioned three Darwinian levels involved in the production of creative human thought. First, there is the biological evolution of genes that has designed the human brain. Second, there is an evolution of neural connections that organizes the brain during Hebbian learning. Third, there is an evolution of looping neural activity among various hierarchically organized circuits, which facilitates the process of thinking. Now, I want to explore a fourth level of evolution operating above the three I just mentioned. It is the evolution of culture and technology occurring among many minds at the level of a society.

We previously characterized the third level of Darwinian process – the iterated looping of neural activity – as a serial style of evolution. It replicates patterns in time, not space. Recall that most such patterns of activity last only long enough to be evaluated. Now, recognize that it would be a much faster process if it could proceed in a parallel fashion. Well, indeed, it can and it does. The fourth level of cognitive evolution achieves a parallel architecture from many serial processes, in many individual minds, all thinking about the same things, in parallel. The mind of society is indeed such a parallel process of cognition – a fourth-level process of cognitive evolution built on top of the lower three Darwinian levels. We can see this more clearly by taking a more comprehensive perspective on the evolution of language.

The evolutionary purpose of human language – its adaptive advantage – is the facilitation of cooperation among those individuals who speak and understand it. Individual activities can be coordinated through language toward the beneficial fitness of an entity at a higher level – the cooperative group. And, when the group thrives, the individual members in the group thrive. Such is the basis for a cooperative morality. And language, then, is the primary facilitator of moral behavior – civilized behavior. It is what allows our serial minds to become synchronized, in parallel fashion, so that we think about roughly the same things. Synchronized thinking enables a form of cooperation toward enhanced replicator evolution through parallel memetic exploration. Such parallel thinking will inevitably discover ever deeper forms of synergistically cooperative activities.

For every opportunity that we ever have to cooperate with other people, we automatically engage in a process of trying to predict the likely outcome of the proposed cooperation in order to assess whether it is worth doing. Cooperation always requires present sacrifice by at least some of the participants in exchange for the mutual future benefit of all participants. The mental simulation of results is the only means by which the sacrificing people can assess the expected value of the anticipated future benefits. The elaborate styles of cooperation among modern humans (especially within and among various businesses) are only possible due to the individual mental means for simulating and evaluating the outcomes of those cooperative efforts. Such mental simulation underlies a capitalistic society's efficient resource allocation through well-considered investments in equity markets. Indeed, mental simulation underlies all modern forms of cooperation.

It is not just a coincidence that societies now facilitate cooperation better than they ever have in the past. Societies have evolved to be that way. But how can this be the case? Societies certainly don't evolve as people do. Societies aren't reproduced in the manner that people are, and whole societies don't die from selection pressures often enough to cause them to evolve. There must be some other sort of mechanism that allows societies to evolve. So, what are the replicators that describe societies and how do they evolve?

When hominids first started to gather into groups, some of those small groups were surely eliminated, along with their cultural behaviors and beliefs, by other groups having better cultural behaviors and beliefs, which likely encouraged more cooperation. In that manner, early cultures probably evolved by group selection. But there must have come a time when societal elimination became rather rare. And today, even without the natural selection of whole societies, cultural memes are still able to evolve. They do so within a virtual environment in the minds of their most influential people. The important point is that societies don't need to physically die in order to evolve. To better see how this is possible, we need to first understand the replicators that define a society.

Whereas primitive societies were defined mostly by memes of religion, modern societies are now defined by memes of government – primarily, their laws and the resulting behaviors of commerce. Societies are defined by memes that create environments of cooperation, which then allow memes of technology to flourish. Cultural memes define, essentially, what a society is and how it functions. Technology memes define what a society is capable of doing. So, societies are defined by their memes, and are merely facilitated by the phenotypes of genes.

From where do a society's memes originate? To answer that, let us contemplate our democratic society. Among the many ideas that get into the brains of our politicians, the ones that cause the simulators running in those brains to suggest a likely path to re-election are the ones that survive. Why? Because politicians hold the belief that getting re-elected is a crucial component to maximizing their long-term happiness. Among the various surviving ideas held by various politicians, the ones that are ultimately selected by a democratic society are the ones that have caused the simulators running in the brains of voters to predict a likely path to greater future happiness for them, the voters.

Among the many ideas that get into the brains of business managers, the ones that cause the simulators running in those brains to suggest a clear path to profits are the ones that survive. Why? Because business managers believe that profits are critical to their happiness. Among the various products that are produced by various businesses, the ones that are ultimately selected by the system of free commerce are the ones that have caused the simulators in the brains of buyers to predict a likely path to greater future happiness for them, the buyers.

It is the collection of simulators running in the brains of society's most influential people that determines the various directions in which the memes of society are able to evolve. And it is the collection of simulators running in the minds of voters and buyers that ultimately selects among those various directions. So, a democratic society's ideas originate in individual minds, but evolve through collective simulation and evaluation.

To illustrate the idea that societies can evolve within some sort of a collective virtual environment, let's consider an example. Imagine a tribe of humans, a long time ago, in which the tribe's leader overhears threats of a pending attack by another tribe in the next valley. He imagines the attack scenario unfolding in his 'mind's eye'. He visualizes a band of warriors one day emerging from the darkness of the forest, screaming and yelling with weapons raised above their heads as they cross the intervening meadow. He imagines the terror that would paralyze the actions of everyone in the community. He actually 'sees' in his imagination all of his compatriots looking toward the approaching invaders with wide eyes and gaping mouths. He simulates, within his thoughts, the devastating details of a surprise attack. His brain automatically mutates those thoughts by combining them with other concepts he has learned. He briefly considers the idea of stationing guards in the surrounding meadow, but rejects that idea as he imagines that those guards would be no obstacle to a surprise invasion of heavily armed men. Again, he 'sees' in his imagination the guards being slaughtered.

One day, while tending his chickens, he realizes that they are protected from the attack of predators by the fence that surrounds them. His brain automatically detects the similarities between the plight of his tribe and of his chickens. An idea suddenly occurs to him – his community needs a wall around it. He knows that such a wall can't be built by a single man, it requires a cooperating society to build it. Thus, the idea is only feasible when conceived by the mind of a society. After telling the idea of a wall to other leaders in the community, they each visualize, in their respective imaginations, the effect of a wall in preventing the surprise element of an attack, and they each recommend improvements in the design of the wall based on their own internal simulations.

It should now be obvious from my example that the society *does* have to die in order for it to evolve. It dies over and over again, but only in the simulated events that occur within the minds of its leaders (and, in the case of a democratic society, in the minds of its voters). Austrian philosopher Sir Karl Popper captured this theme when he said that our human ability for mentally simulating the outcomes of our ideas "permits our hypotheses to die in our stead."

It is rare these days that an unfit society is eliminated by a competitive society that is fitter; although, the organizational memes of the former

Taliban society of Afghanistan seem to have suffered just such a fate. Realize that a society is defined, not by its people, but by the imposition on its people of its organizational memes.

It is now more often the case that the memes of a free and democratic society evolve through mental simulations performed by all the voting members. Such memes are selected on the basis of their consequent appeal to the emotions of those many members. Do new governmental laws (memes) make them feel safer? Are the new laws likely to satisfy their lusts and their hungers? The simulated answers to those questions, within the minds of a democratic society's members, represent the selection forces for the evolution of the respective societal memes.

It is only through the memetic perspective on culture that we are able to view the society as a living entity unto itself, constantly adapting through evolutionary forces operating within the minds of its influential members. But, since those evolutionary selection forces are shaped by causal expectations of happiness, and since happiness is naturally defined by genes, and because causality is a consequence of nature, then we may conclude that the forces of nature ultimately determine how cultural memes will probabilistically evolve. Of course, we've known since chapter four that the forces of nature determine the direction of evolution for all replicators, at least in the statistical long-term. And we should have suspected as much if indeed the brain is deterministic, as was claimed in chapter one.

We should now easily recognize nature's primary over-riding principle of selection: A replicator's fitness is determined by its ability to cooperate with other replicators in the pursuit of synergistic benefit toward their mutual perpetuation. This supports the concept that perpetuated synergy is the essence of life. It should be no surprise that human brains are organized so as to recognize and use the synergistic effects of memetic cooperation. Indeed, our human brains were originally designed to look for combinations of efforts that would benefit the perpetuation of our genes. But, our ideas have become such a critical part of us that we are now built by our genes *and our memes* to find combinations of ideas that will perpetuate them both, including all the organizational memes of our society. It seems we individuals have become naturally integrated into a greater living entity.

Over the long term, fitness for a given society is determined by how well that society encourages cooperation among its members. Only a highly cooperative society is capable of such advanced technology as we enjoy in the United States. The societal memes of the U. S. (including those of its free-trading partners) are highly conducive to the process of discovering and developing useful memes of technology. Patent laws and capital markets are structured so as to encourage the creative pursuit of technology. Libraries, journals and Internet websites are available as means for facilitating the flow of ideas from mind to mind to mind. Consequently, memes of culture and technology are constantly evolving, and are doing so ever faster.

Large cooperative societies are statistically more capable of discovering useful meme combinations than are small societies, or, especially, individuals. An invention may emerge in the mind of a single individual, but it would be unlikely to do so without that individual having been strongly influenced by many others in its society. Indeed, the discovery of new technology now comes primarily in the form of many people exploring, in parallel, the vast hyperspace of possible memetic combinations. It should not be surprising that metropolitan areas tend to file numbers of patents in relation to the sizes of their populations. This style of brute-force cooperation toward memetic discovery is an additional form of societal benefit over and above the more traditional forms of economic benefit resulting from societal cooperation (division of labor, specialization, economies of scale, and diffusion of risk). Once an obscure and valuable idea is discovered, all in the society can benefit from it.

The cooperation facilitated by language produces a synergistic mind of society that exceeds even the sum of all intelligence over all the minds of its members. The society takes on a mind of its own attributable to the simple notion of replicator combination and re-combination, which naturally and automatically occurs across all cooperating and communicating human minds. The societal mind is simply a parallel implementation of the serial evolutionary process that goes on in an individual mind. As such, it is far more creative than an individual mind.

The societal mind, as creative as it is, will take a great leap forward when computers become capable of human-like creativity. Indeed, the society that successfully advances automated intelligence the farthest and the quickest will command the world's precious resources. In the history of the world, allocation of *scarce and necessary* resources has rarely been a matter of diplomacy or fairness, it has usually been a result of the fit overpowering the unfit.

Evolution is no longer operating at the level of genes and the individuals they create, so much as it is operating at the level of memes and the technological societies they create. We would be foolish to think that a future society, faced with the prospect of diminishing survival resources and the consequence of extinction, won't fight to the death for control of such precious resources as food, fresh water and energy, if its survival depends on it. It seems to be a matter of destiny that the fates of all societies will eventually be determined by the intellectual power of their computers. Using intelligence for enhancing the survival and perpetuation of replicators, now including memetic as well as genetic, has always been and will forever be the ultimate evolutionary purpose for intelligence.

Chapter 9. **The Computable Mind**

All evidence suggests that our human minds are nothing more than brains, and brains are nothing more than computational machines. Our thoughts now appear to be mere patterns of neural activity tracing and re-tracing hierarchical representations of real-world relationships, creatively evolving toward plans that promise to satisfy whatever goals have been established by the biological evolution of our genes. Our brains typically guide our behaviors in ways that are similar to successful behaviors we have witnessed by others. And, once in a while, an active mind will evolutionarily discover a new beneficial combination of behaviors that other minds will assess to be beneficial and will surely mimic.

The foundation has been laid for describing the human mind as being purely computational. We now find ourselves in a position to describe, computationally, the inferential cognitive processes of *induction* and *deduction*. And, we are now sufficiently prepared for appreciating what it means, from a computational perspective, to *understand* something. So, those will be our endeavors in this chapter. Once we have those concepts under our belts, we can then begin to contemplate computational models for intuitive reasoning and for processing natural language.

Until now, the text has been fairly rigorous and well-focused in developing the critical aspects of the book. From here on out, I will lengthen the tether a bit on the topics I discuss. I feel justified in occasionally mentioning some loosely related topics that aren't really critical to the theme of the book but are somewhat interesting nonetheless. Yet, the next topic describes an inferential process that is absolutely crucial to intelligence.

The Process of Induction

Induction is the process of grouping patterns into various classes based on their similar characteristics, and classifying incoming sensory patterns on the basis of already-established groups. It is where intelligence begins. The evolutionary value of induction is not hard to see. For instance, we may logically suppose that a very important function of the primitive evolving mind was to recognize the difference between food and non-food on the basis of patterns sensed in the environment. Perhaps those were the very first categories into which environmental objects were mindfully separated.

Regarding our modern brains, as well, induction is a critical cognitive component for intelligently interacting with environmental circumstances. As our environments change, we need to sense those changes and deal with them appropriately. One manner of doing so is by recognizing new environmental situations – patterns – as *similar* to previous ones with which we've already witnessed ourselves or others successfully coping. It is the notion of similarity between sensed patterns that makes induction so intriguing, from a computational perspective. We have already discussed the computational metric used by the brain to loosely quantify similarity among patterns. We'll elaborate on it soon.

Once a situation is inductively recognized as similar to some remembered situation, the manner of response depends on the level of intelligence. *Primitive* intelligence involves mimicking whatever past actions turned out to be appropriate in previous similar situations. *Advanced* intelligence involves the application of an evolutionary process so as to produce a plan that is likely to yield future benefit, given the circumstances that are inductively sensed. That evolutionary process creates many different combinations of activities that, when simulated on the basis of expected causality, can be assessed in terms of the amount of value they are each likely to produce.

Primitive induction is performed by primitive life on the basis of very simple and easily recognizable characteristics. But human induction, on the other hand, involves classification on the basis of very abstract characteristics. For example, what is it that makes a specific character of text, say, an 'A', recognizable in any of a thousand different fonts? It is not just the relationships between lines and angles but the relationships between relationships between lines and angles. We humans can easily read text printed in a brand new style of font, never seen before, so long as the font designer has faithfully maintained all the Nth-order abstract relationships that somehow define the text characters.

The process of induction is what causes some people to believe they see an image of the Virgin Mary on a grilled cheese sandwich, or in the patina of dirt on a plate-glass window[8]. Whatever abstract relationships happen to define the appearance of the Virgin Mary, we should expect them to occasionally emerge from randomly generated patterns, given enough opportunities ... even on a grilled cheese sandwich.

The process of induction is why we see objects in the blobs of ink that represent a Rorschach test. It is how we categorize the objects that we see into various classes, and how we draw inferences about the nature and functionality of those objects by virtue of the classes to which they appear to belong. We have already seen hints that induction springs directly from Hawkins' memory-prediction model.

The fact that various patterns can be similar in varying degrees leads us to suppose a continuous hyperspace in which patterns exist. Indeed, given any two similar patterns, there always exists some pattern in between the two that is more similar to each of the given patterns than the given patterns are to each other. Unfortunately, this continuous hyperspace of patterns leads to an inherent ambiguity in our various perceptions of the world. As if quantum uncertainty were manifesting itself at a higher level, the boundaries of classes of patterns are necessarily fuzzy. Like trying to establish a meaningful boundary between the colors red and orange, we find that the continuous domain of perception has no clear, absolute boundaries. Nevertheless, we intuitively define statistically useful boundaries through the automatic process of induction.

All knowledge is necessarily built on a foundation of concepts that are ultimately not provable. The ill-defined bricks of that foundation are the many inductive generalizations that we've made. For example, we assume that all ravens are black, even though no one has ever seen every single raven. We generalize in this manner because it is the practical thing to do. It yields statistical benefit.

After seeing a large number of ravens, all of which are black, we automatically draw the inductive inference – the generalization – that all ravens are black. Such inferences are not always valid, but they are valuable nonetheless. They are practical tools of intelligence that can probabilistically enhance the chance of survival by drawing conclusions that are likely to be true, or mostly true. For example, it is statistically beneficial to assume that all lions are dangerous, even though there may be a few that are actually as harmless as 'pussy cats'.

We feel justified in generalizing across a species, such as lions, but not across a race of humans. Indeed, the act of generalizing across an entire race on the basis of observing just a fraction of its members is what racists do, and is deemed quite inappropriate in that regard. But it is a fact of human nature that we automatically gather inductive statistical data for use in future decision making. Indeed, we all automatically perform unfair inductive inferences. For instance, after the World Trade Center attacks, it became common for air travelers to scan other passengers for Muslim-looking men. The reason we all seem to do it is because nature found it practical. And, from a statistical and practical perspective, we have to admit it is a logical thing to do in an environment of imperfect information.

The inductive process is one of generalizing from specific instances to broad classes. A human mind automatically carves out a region of conceptual space for a particular class of pattern, and then automatically assigns attributes to the entire class based on relevant experiences with just a few instances of the class. Despite the adverse sociological implications of induction, all of science rests upon its shoulders. For instance, scientists operate under the belief that the force of gravity applies to all massive objects, even though no group of scientists has ever checked them all. Ultimately, everything comes down to a matter of definition. So, if we find an object to which gravity does not apply, then, by definition, it is not a massive object. But, what if that same object exhibits inertia, as if it has mass? We must then define a brand new category for it.

At its essence, science is a simple process of gaining enough information to define classes and relationships between those classes such that real things can be properly classified with absolutely no exceptions. So, in a scientific world with perfect information, racism disappears, simply because our having perfect information allows us to classify only bad people as being bad, completely independent of their race. Yet, in the real world, where there can be no such thing as perfect information, there is no way to define a class of objects and be sure that our classification is completely accurate, without checking each and every instance of the class. For example, we simply can't say with certainty that all neutrons have mass, and are therefore affected by gravity, unless we check them all.

The only way we can bridge the gap between the realistic limitation on our knowledge and the practical value of making generalizing assumptions is through the principle known as Ockham's razor: All else being equal, the simplest explanation tends to be the correct one. Indeed, it is much simpler to assume that all neutrons have mass than to suppose that some are possessed with unique characteristics.

Because of our inability to know everything about the universe, we can't really know anything with certainty. Whatever factual explanation you give me, I can conceive of a ridiculously improbable alternative that just may be true. Suppose you tell me that the moon is held in orbit by gravity. I am perfectly justified in replying, in the spirit of a precocious ten-year-old, that it just may be held in orbit by Superman, flying undetected on the far side. The only means you have for disproving my precocious explanation is on the basis of Ockham's razor. We can only *approach* certainty through Ockham's statistical tool.

For similar reasons, I must admit to my religious friends the possibility that the human will may indeed derive from a supernatural spirit that is implanted by God at birth. But such an explanation begs many more-complicated questions: When in evolutionary history did God start implanting spirits? Do dogs have spirits? How about worms or bacteria? Why did God bother with evolution at all? How does a supernatural spirit connect with a physical brain and still maintain its supernatural status?

The simplest explanation is that the human will is a product of physical brain processes alone, and that those brain processes are products of evolution. The salient point is that no sort of intelligence, human or machine, can ever be completely certain of its knowledge. All intelligence must ultimately rely on Ockham's razor, which is only a valid principle in a probabilistic sense.

Philosophers have understood the shortcomings of induction for centuries. While the process of induction allows us to draw valuable inferences from sets of premises, those inferences are not guaranteed to be true, even when the premises are guaranteed to be true. For example, even if it is completely true that I have seen thousands of ravens, and everyone of them was black, it is not necessarily true that all ravens are black, even though such is the conclusion that tends to be drawn.

With a clear understanding that inductive inferences lack any sort of grounded formal logic to guarantee their conclusions, let us realize that they are still critical to intelligence, from a statistical point of view. All human acts of intelligence require the recognition and classification of patterns, and the process of classifying patterns is a matter of induction. Any model of neural architecture must account for induction, and, as it turns out, our current model does so, beautifully.

Mapping Sensory Data to Conceptual Space

If I show you a picture of my dog, it will likely be an image that is at least slightly different from any that you have ever seen before. Yet, you and many others will easily be able to classify the image as that of a dog. So, then, what are the characteristic qualities that allow it to be perceived as a dog? Whatever the dimensions of those characteristics are, they form a sort of *conceptual hyperspace*. There is a region within that grand conceptual space that corresponds to the class 'dog', and any pattern having characteristics that map to a point within that region will be classified as a dog.

The region of conceptual space corresponding to the class of all dogs is further subdivided into classes corresponding to various types of dogs (collie, poodle, spaniel, etc.). At the centers of all those sub-regions are prototypical definitions for those various types of dogs. The prototype of the class poodle, for instance, is what one would draw if one were asked to draw a poodle.

When we talk about a classification hyperspace, we are making an implicit assumption that every possible pattern maps to a unique point in that space. And, more importantly, we are implying a means for assessing similarity between any two patterns simply by measuring the Euclidean distance between their respective points. But, how then might this metric of closeness relate to our proposed neural architecture, which consists of many overlapping relational hierarchies? The answer is revealed by considering how visual patterns map to relational hierarchies for various dogs.

In a given brain the collection of all dog hierarchies, corresponding to every dog ever seen, defines the region of conceptual space corresponding to the class 'dog'. And, likewise, the collection of all hierarchies corresponding to every poodle ever seen defines the sub-region corresponding to the subclass 'poodle'. Indeed, many instances of similar hierarchies, whatever they represent, will map to a cloud of points in conceptual hyperspace. So, a classification region can only be vaguely defined by such a cloud's ambiguous boundary.

Different pictures of dogs will excite different but overlapping cortical hierarchies. The more similar the pictures are, the more overlap there will be between their corresponding hierarchies. Overlap, in this sense, refers to nodes that are shared among hierarchies. Whereas all hierarchies share sensory elements at their root tips, the sharing of nodes at slightly higher levels determines the degree of abstract similarity between any two hierarchies. Similar patterns coming from the senses will map to similar hierarchies, and similar hierarchies are those that share some abstract nodes.

We have loosely defined a measure of similarity between any two independent things in terms of the degree to which their respective hierarchies overlap and thereby share both primitive and abstract components. Thus, if two dogs have similar-looking big and floppy ears, they will share some intermediate-level nodes that correspond to that particular style of ear.

The prototype of any class is defined by whatever nodes are shared by most of the instances of the class. So, a prototypical poodle will be defined by whatever nodes are shared by most of the poodles ever seen. Consider the flow of neural activity that happens when someone mentions a poodle. The corresponding image is conjured by neural activity going to the topmost node of the hierarchy corresponding to the class of all poodles, which in turn sends activity downward to all hierarchies of every poodle ever seen and remembered. Those hierarchies in turn send activity downward to all the abstract relationships that define all those poodles. The nodes that are most commonly shared by the hierarchies representing all poodles ever seen are the ones that are likely to be most active, simply because they receive downward-flowing excitation from many different higher sources. Those active nodes define the imagined visualization of a prototypical poodle.

The way that classes of things first become established is really quite simple. When a very young child sees an instance of something that others call a dog, the characteristic qualities of that dog map to a specific point in the child's conceptual space – to a specific hierarchy in its cortex. The fact that others call it a dog allows the child to assign a label to the hierarchy, which initially defines the class of dog. So, children merely mimic from others the assignments of labels to various classes.

When an older child sees a new dog, the child no longer needs to know that others call it a dog. The child is now able to automatically make that assumption on the basis of its already-established region of conceptual space corresponding to the concept of a dog. As the child gathers more and more points in that region of conceptual space from seeing more and more dogs, the conceptual region corresponding to dogs becomes more and more clearly defined in its mind.

All day long, every day of our lives, our brains build up these conceptual regions. They are defined by all our experiences over the courses of our lives. And, using these regions, we are easily able to classify patterns that are similar but not identical to others we have seen before. The classification of a brand new pattern happens, quite simply, on the basis of overlap between its resulting neural activity and existing hierarchies.

When a brain is confronted with a new visual pattern, its task is to map the characteristic qualities to a point in conceptual space, and thereby determine what class of object the pattern represents. The inductive process yields an undefined result when the point maps to a boundary between two different classes of object. For example, suppose I were to show you a picture of an animal that appears to be midway between a dog and a cat. It will excite a cortical hierarchy that partially overlaps both the hierarchy for a prototypical dog, and also, the hierarchy for a prototypical cat. The resulting classification will be indefinite, yet still useful for metaphorically assigning attributes to it on the basis of similarity to both dog and cat. You might logically suspect that the odd-looking animal makes a sound somewhere between a bark and a meow, because the auditory hierarchies corresponding to both a prototypical bark and a prototypical meow will both be excited, and their overlapping regions will therefore be doubly excited.

All our experiences affect in subtle ways our conscious views of reality. We all perceive the world in terms of patterns to which we've previously been exposed – visual patterns, audio patterns, olfactory patterns, patterns of behavior, and patterns of patterns of patterns. As Edelman and Tononi make clear in *A Universe of Consciousness* (2000, p.174), our tastes and our sensations evolve over the histories of our exposures to new things. "For example, until we gather sufficient experience, different wines taste more or less the same. But soon, their taste will become associated with strikingly different qualia. Clearly, where there was only the ability to discriminate wine from water, there is now the ability to discriminate reds from whites, and Cabernets from Pinots." As I previously argued, the conscious ability to discriminate among the many sorts of things we sense grows out of a growing history of exposures.

Repeated exposure to the same pattern makes the corresponding neural detectors more selective and more sensitive. Just as repeated exposures to various wines make us more able to discriminate between them, so do repeated exposures to, say, constellation patterns allow us to more easily pick them out of the nighttime sky. I remember as a kid being incredulous at the idea that the stars form pictures. I saw no sort of man in the constellation Orion. But now, as an adult, I see the shapes much more vividly.

I previously suggested that induction is where intelligence begins. Indeed, the process of induction gives our brains all the information they ever have about the world. Fundamentally, it is induction that allows us to discriminate between a tree and a pretty girl. And, as we'll soon see, it is induction that forms the basis for the human ability to understand various sorts of abstract concepts.

Semantic Understanding

What is it that constitutes an understanding of some new situation? And, what causes the accompanying *feeling of understanding*? I endeavor to show that, with regard to some new situation, understanding occurs whenever the process of induction successfully maps the relationships among patterns of the new situation into some familiar class, which also happens to include a link to the knowledge of how to successfully deal with it. And, when understanding occurs, a *belief* in a conscious *feeling* of understanding automatically becomes active. Understanding, then, is mostly a matter of recognition, or successful induction.

Hawkins suggests that to understand something is to be able to make predictions about it. But the only way to make predictions about something is to relate it to previous experience. And the only way to relate something to previous experience is to recognize it as being similar to something seen before. So, while I completely agree with Hawkins' definition, I believe it is more fundamental to describe understanding as mostly a matter of inductive recognition.

I understand something when my brain recognizes a way of successfully dealing with it. For instance, I understand the expression '18 + 44' because I recognize all the mathematical symbols, and I know exactly what to do when a '+' stands between two numbers. I don't understand Chinese writing because I don't recognize any of the hànzi characters. And, even if I did recognize one of the characters – say, a familiar tattoo on some movie star's forearm – I still wouldn't recognize a manner of dealing with it. I wouldn't recognize to what in the world it relates.

We can gain some more insight into what it means to understand something by reconsidering our previous discussion of Searle's Chinese Room (from chapter six). Recall that the English-speaking man inside does not understand the Chinese symbols on the slips of paper he receives through a small window, yet he is able to process them in a way that makes the entire room appear as if it understands Chinese. He processes the symbols by looking them up in various books of tables, following the corresponding English instructions to generate a response.

The concept of understanding something is easily illustrated, within the context of Searle's Chinese Room, by considering what it means to *not* understand something. If the man in the room receives a slip of paper with symbols on it that have no matching entries in his translation tables, then he has no means for processing it. And, consequently, the room appears to *not* understand the writing on the slip of paper, because the room gives no response.

By modeling feelings of understanding in the same mechanistic manner as we have learned to model all feelings of consciousness – as nothing more than beliefs that become true under certain conditions – then we can easily give Searle's Chinese room a *feeling* of understanding when it successfully interprets and responds to the incoming symbols on the slip of paper. The man simply presses a button that lights up a sign with the expression 'understood'. If he gets a slip of paper for which there are no matching entries in his lookup tables, he presses another button that lights up a sign with the expression 'huh?'.

If the man in the Chinese room gets a slip of paper that asks him whether he understood the last message (unrecognizable to him, because this message is likewise written in Chinese), the lookup tables will direct him to query the room's belief by looking to see which sign is lit. By such 'self reflection' the room and its lookup tables are able to determine whether to respond with the Chinese symbols for 'yes' or for 'no'. The room doesn't actually *feel* a feeling of understanding, but, from all perspectives including its own, it believes it does. Whenever the 'understood' sign is lit, the room believes it has a feeling of understanding. Whenever the 'huh?' sign is lit, the room believes it has a feeling of being perplexed. I am inclined to believe the same must be true of humans: When an incoming sensory pattern sufficiently matches the detection pattern of some input relational hierarchy so as to cause neural activity to reach a high-level node, which in turn produces a strong downward-flow of activity confidently specifying some imagined response, then there automatically occurs a corresponding belief in a feeling of 'understanding'.

While mere recognition is a critical first step in understanding, there is a much deeper level of understanding that can only occur when *value* is involved. Recall my earlier suggestion (in chapter one) that data only becomes information when it provides value to those who know about it. It should not be surprising, then, that at least a small part of the feeling we often get from understanding a piece of information relates to simply finding the value in it. For instance, if someone were to hand me a slip of paper with the words "grass is green" written on it, I would be dumbfounded. Even though I recognize all the symbolic tokens and the semantic content in that message, where is the value? Until computers are programmed to value some things over others – as humans value having sex and eating chocolate over stubbing their toes and eating wood – computers will never understand certain things the way we humans do. The question then becomes: Can we artificially give computers such values? Absolutely we can, if we want to.

Let us analogize the concept of 'value' within the context of Searle's Chinese Room. Suppose that, occasionally, a submitter of questions wishes to express his gratitude to the man in the room. He writes "thank you" (in Chinese) on a slip of paper, smiles, and hands it through the window, then reaches into his pocket for a few pieces of hard candy, which he also hands through the window. Before long, the man in the room will come to recognize that certain Chinese symbols (for "thank you") are soon followed by a delicious treat. The 'room' now connects value with certain characters, and a deeper level of understanding then occurs.

It turns out that we can only fully understand symbolic tokens, such as words or Chinese symbols, by relating them to real-world objects and actions, because only real-word objects and actions can have value. Terrence Deacon touches on this concept when he says the following:

> Symbolic reference is grounded by the relationships between a system of token-token relationships and a system of token-object relationships, but the walls of the Chinese Room make symbolic interpretation impossible because they make it impossible for the man to discover any relationships between the token-token system in the book he has access to and the systems of token-object and object-object relationships that are unavailable to him. … No set of preprogrammed algorithms can smuggle symbolic reference into Searle's Chinese Room, because symbolic reference cannot be solely inside.

> **– Terrence W. Deacon**, *The Symbolic Species* (1997, pp. 446-447)

By handing the man some candy, a real-world symbolic reference – a semantic reference – is smuggled into the room, which allows the man to assign some level of meaning and value to certain Chinese characters.

Such is the nature of teaching language to a child. By showing a child an instance of something in the real world, and pronouncing the corresponding word, we smuggle a symbolic reference into its brain – we establish a relationship between the visual image of some object in the real world and the corresponding audible word. We have now established three levels to understanding, involving: (1) the simple recognition of tokens, (2) the recognition of relationships between tokens and real-word objects or actions, and (3) the recognition of value in the objects and actions referenced by the tokens. I can illustrate these three levels of understanding by sharing a personal experience of mine.

I used to read short detective stories to my daughter at bedtime when she was very young. In one of them, the primary suspect claimed to have been hiding under water during the crime of which he was accused. He claimed to have been breathing through a six-foot length of bamboo. His story was proved false by a detective who calculated the volume of air in the bamboo to be greater than the volume of air in a typical person's lungs. Now, do you *understand* how the detective knew the alibi was false?

If you understand, it is because you have inductively recognized the situation of breathing through a long tube as being similar to my previous discussion regarding the evolutionary development of the long necks of giraffes. Recall, as I previously mentioned, that the emergence of long-necked giraffes required the co-adaptation of large lung size, because, if a tube through which one breathes is of larger volume than one's lungs, then one will inhale the same air that was just previously exhaled. Since giraffes breathe through long necks, they need larger lungs to compensate. Through induction, my mind was able to recognize the abstract similarity between the patterns of breathing through a long bamboo tube and breathing through a long neck.

A full understanding of this story requires the recognition of the three primary ingredients I previously recited as necessary for complete understanding: tokens, relationships and values. First, one must recognize and be familiar with all the important words (tokens) in the story. Second, one must recognize the relationships between various words and the things they represent. For example, if one has never before seen the object to which the word 'bamboo' refers, one may not realize that it is hollow. Third, one must recognize the value in learning the causal relationship capturing the fact that one cannot survive while breathing through a long hollow tube.

Ever since reading that detective story, many years ago, I carried around the associated causal meme in my brain. It only recently became excited by an abstractly similar pattern, in my Nth-order relational space, when I happened to consider the evolution of the giraffe's peculiarly long neck. Insight always results from an inductive process of noticing abstract similarity between whatever one is contemplating and some other concept with which one is already familiar.

The process of induction is always the first step in general problem solving. When a brain is faced with some problem to solve, induction is automatically performed on the parameters of the problem to see if there are any existing means for dealing with it. Is it a math problem? If so, the recognition of mathematical symbols will induce an automatic plan of action, along with a feeling of understanding.

Analysis of the problem to be solved is certainly a big part of intelligence. And, problems generally fall into one of many classes, based on the patterns of their parameters. If the inductive process recognizes a given problem as a familiar one, then the procedure for solving it is clear, and a belief in a feeling of understanding automatically results. Otherwise, the analysis procedure resolves to an evolutionary process of trying and evaluating various combinations of known approaches to all similar classes of problems.

We are methodically describing many aspects of human thought in computational terms. Let us now consider the very dynamic nature of how our memories are encoded, from a computational point of view.

Dynamic Encoding

We like to think of the world in terms of objects and their characteristics. Perhaps our propensities in that direction grew out of an evolutionary advantage from possessing and coveting *things*, such as hoards of food, weapons and tools. But, while we tend to think of many objects in terms of their static characteristics, our brains encode them completely in terms of their dynamic characteristics.

Think of a baby's first encounter with a cube-shaped block. As neural firings come into the brain from the eyes, they describe a visual pattern that is, for the moment, static and uninteresting. The baby doesn't learn very much until the block is touched, caressed, turned, brought closer, and then thrown back down to the floor. As the block is turned, the baby's mind builds a huge library of visual images all related to the same sort of thing – images of a cube-shaped block from all angles. Every viewing angle leaves an impression in the form of a specific connection hierarchy, but only if it is sufficiently different from all pre-existing hierarchies (repeated instances of the exact same viewing angle map to the same hierarchy). The enormous amount of data captured and stored is only possible due to the very efficient means for encoding images, which involves the heavy sharing of abstract, visual primitives.

Due to the resonance of short-term memory, all the viewing angles become linked together at a higher level into a single concept corresponding to a block. From any viewing angle, the perception of a block will inductively map to a point within the region of conceptual space that corresponds to the class of all blocks. And, the dynamic act of turning a block will trace some path through that same region. If you are having trouble imagining how every single object can have its own region in conceptual space, it is because we all think of space in terms of just three dimensions. But, conceptual space includes many additional dimensions of abstract relationships, simultaneously. Don't bother trying to visualize the dimensions of conceptual space and how they might map into real space – no such mapping can possibly exist[9].

The baby's brain doesn't store the static spatial dimensions of a block, but rather, it stores the dynamic experience of handling a block. The brain knows nothing of coordinates. It knows only about neural patterns evoked by interacting with the block, and those patterns are likely to be very dynamic. There is a huge difference between the way in which computer programmers tend to store knowledge in computer memory and the way in which the human brain stores knowledge. The difference is something like the difference between a still-picture and a movie.

Modern computer programmers tend to encode their data the easy way. They encode static characteristics of things and then program in mathematical or logical rules by which those static representations can change. There might be a rule that says 'if an object has flat surfaces, then it slides when pushed', and, 'if an object has a round surface, then it rolls when pushed'. This is fine so long as we capture every possible rule corresponding to every possible scenario that can be applied to our static objects. That is a tall task. And it often turns out that the rules tend to have many exceptions, all of which need to be captured as well.

Nothing about a static representation of either a block or a ball provides any information regarding how they each will react to being pushed. Yet every child quickly learns that balls roll and blocks don't. A child's brain automatically collects many dynamic 'movies' of blocks and balls being pushed. When a baby contemplates pushing either a block or a ball, its brain automatically searches for a similar, remembered movie snippet. If a close match is found, a feeling of understanding occurs.

If we ever want computers to be capable of doing the kinds of things that human minds can do, then we had better change the way they store knowledge from static representations to dynamic ones. In order to think like humans, computers will need to collect and remember every experience, dynamically, then look for matches to existing conditions for indications of how the future will likely unfold. Unfortunately, this too is a tall task.

Luckily, overlapping hierarchies naturally produce an abstract form of data compression that makes the task tractable. As we humans collect our experiences, our brains automatically compile them down to patterns that are abstract representations of things. A computer need not remember every instance of a block sliding or a ball rolling. Similar instances of dynamic activities, such as various memories of balls rolling, will merely reinforce each other by mapping to the same causal concepts in abstract relational space. They merge together in an abstract sense. Thus, our dynamic mental movies don't exist in an image format, but rather, in the abstract format of conceptual relationships, which are linked together in ways that make them dynamic.

Whatever abstract characteristics differentiate blocks from balls, the important characteristics are mostly dynamic in nature. And our inductive processes enable us to categorize things by their dynamic characteristics as well as their static characteristics. For example, they allow us to determine the likelihood of some new type of object either sliding or rolling on the basis of its abstract similarity to things about which we already know.

The Logic of Deduction

Whereas induction is a process of classification on the basis of things already known, deduction is a process of creatively exploring that which can be known, by combining things that are already known. We'll now contemplate what it means to draw an inductive inference, and we'll later examine the deductive logic often used by computer scientists in the field of artificial intelligence. We'll find that common techniques for programming the process of deduction bear significant similarity to the previously developed concept of achieving creativity through an evolutionary process of combining causal memes.

The simple value of deduction lies completely in its ability to reveal implicit information as explicit. For example, suppose we know it is true that: (1) *All good chess players are smart*, and (2) *Jimmy is a good chess player*. Those two facts, by virtue of their common relationship to the concept of a 'chess player', actually express more knowledge than they explicitly state. A third fact, (3) *Jimmy is smart*, is already implicitly stated, but it takes a deductive inference to make it explicit.

Unlike inferences derived through induction, which can only be valid in a probabilistic sense, deductive inferences produce conclusions that *do* follow from their premises, absolutely. Indeed, the process of deduction lays the foundation for rigorous systems of logic. If we can know with complete certainty the truthfulness of a given set of facts, then we can maintain complete certainty in the truthfulness of any logical deductions produced from that set. The set of explicitly stated facts can thereby grow in number. Such is the rigorous underpinning for logical creativity. The specific means for creating new facts involves an exhaustive search for existing facts having complementary interfaces, just like the combining of causal memes.

While logical deduction can indeed be a perfectly rigorous system of rules for revealing implicit information as explicit, the value of the rigor is somewhat mitigated by the fact that deductive logic necessarily operates on relationships that can only be attained through the probabilistic process of induction. For instance, the statement *all good chess players are smart* is an induction based on experience with many chess players, all of whom happened to have been smart. But we can't claim to have checked them all. And the statement *Jimmy is a good chess player* can never be an absolute fact, because even though Jimmy might have won his last hundred chess matches, it is possible that they were all flukes.

We are left with the conclusion that absolute truth is a slippery concept. We have already discussed how quantum uncertainty prevents us from knowing any perfect truth whatsoever about the state of the world. Now we have even more evidence that perfect truth is unattainable. But, we have to do the best we can with what we have. And it seems to be the case that we *can* know things probabilistically, to a very high degree of confidence. So, let us simply keep in mind that our facts can only be probabilistically true.

Philosopher Karl Popper recognized the unattainable nature of truth, and that the consequent uncertainty necessarily implies an evolving nature of knowledge. Here's how the argument goes: No amount of verification can ever completely confirm a theory as being true. Yet, only a single instance of falsification can decisively prove it false. So, Popper proposed falsifiability as the criterion for validity of knowledge. Instead of assuming a theory is false until proven true, let us assume it is true until falsified. Thanks to Popper's insights, it has become widely recognized among scientists and philosophers that any theory that is not falsifiable, such as, say, the existence of ghosts, is not a scientific theory at all. It is reduced to mere pseudo-science because of its inability to be tested for falsification.

Because of the guaranteed inability to prove any theory absolutely true, the advancement of knowledge, according to Popper, must be considered as an evolutionary process. Given some problem situation, competing theories are subjected to rigorous attempts at falsification. Such a process is precisely analogous to the role natural selection plays in biological evolution. Theories that survive the competitive and selective process of falsification are not more true, but rather more 'fit' – they fit better with empirical data. Just as the fitness of a biological species does not guarantee its continued survival, neither does the fitness of a theory guarantee an immunity to eventual falsification. Just as biological species evolve, so must causal theories evolve. It seems that Popper advocated an evolutionary theory of knowledge long before Dawkins proposed his evolutionary theory of memetics.

For a clear illustration of Popper's reasoning, consider Newtonian mechanics. Its associated theories were believed to be absolutely and precisely true. But then along came Einstein and his theory of relativity, which proved Newtonian mechanics to be incorrect. So, is Einstein's theory of relativity now to be considered absolutely true? No, it is only ever logical to consider the prevailing theory to be the 'fittest' rather than to be absolutely 'true'. This is necessarily the case because all theories depend on the imperfect process of induction.

Just as the fitness of any biological species is determined by laws of nature, so is the fitness of any combination of ideas similarly determined. And, just as biological evolution discovers ever more forms of synergy resulting from various combinations of atoms, so does the evolution of knowledge discover ever more forms of synergy resulting from various deductive combinations of ideas, which are themselves ultimately built up from fundamental facts of nature.

Of course, fundamental facts of nature must be discovered empirically. For instance, Galileo needed to drop a couple of objects from a high tower before he could know for sure that they would fall at the same rate, independent of their different weights. And, no one could have known that the speed of light is a constant, independent of the measuring frame of reference, until Michelson and Morley did the experiment in 1887 that revealed it. But once that fact was known, Einstein's special theory of relativity immediately became implicitly known as well. Yet, it took Einstein's deductive reasoning to combine the complementary facts in a way that would reveal the implicit information as explicit.

Einstein imagined how light, emerging from a train traveling at or near the speed of light, would appear to a person on the train or to a man standing on the platform of a train station as the train goes by. He discovered he could only reconcile all the perspectives by demanding that the passage of time be relative to each observer and dependent on an observer's frame of reference. This further undermines the notion of absolute truth with respect to time and space. Time, which was once thought to be an absolute for everyone, now turns out to be relative to one's frame of reference. But, despite the guaranteed inability to achieve perfect and universal truth, we are certainly able to increase our level of confidence with respect to how things work in any given frame of reference. We do so by inferentially revealing implicit information as explicit.

One way of turning implicit information into explicit information is by simply proposing various theories and then seeing which theory best fits all the empirical facts. This is certainly consistent with Popper's view of theoretical 'fitness' over theoretical 'truth'. Sometimes it takes a wild theory to get the job done, as was certainly true in Einstein's case. Einstein must have had an abstract meme, regarding something like elasticity, that allowed him to abstractly conceptualize the stretching of space and time. But, he needed to mutate the mathematical representation of elastic reality over and over again in various ways to get it to line up with empirical evidence. There is often a reverse-engineering process to get started, but then forward evolution takes over. Recall that reverse engineering is simply the automatic discovery of a causal meme whose output matches the goal.

Now that we understand how creativity can emerge from a deductive process that turns implicit information into explicit information, let us contemplate how such a deductive process can be automated using present-day computers. I'll start by discussing the commonly accepted techniques, among computer scientists, for automating the process of deductive reasoning. I'll show the inherent weaknesses of those techniques by contrasting them with the expressive and functional power of relational hierarchies.

It is common for computer scientists engaged in the pursuit of artificial intelligence to use relationships as the basis for their model of knowledge. They speculate that the world can be completely described in terms of a huge list of relationships, and that there are algorithmic routines that can be automatically applied to the list in a way that simulates the process of human intelligence. Good for them. They are on the right track, given our new understanding of how the human brain apparently captures real-world relationships through Hebbian learning.

Let us consider an example of how such a list of relationships can be manipulated by deductive logic. Suppose, in a long list of relationships, one of them declares that *all kids love candy*, and another that *Jimmy is a kid*. It seems straightforward to derive the obvious inference that *Jimmy loves candy*. We can indeed automatically derive such deductive inferences by expressing the given factual relationships in a very definite shorthand notation that lends itself to automatic manipulation by algorithmic processes. There are in fact several styles of shorthand notation commonly used by computer programmers for expressing information regarding objects and their relationships. A popular class of such ontological notations is known as *first order predicate logic*.

Given a list of valid statements expressed in first order predicate logic, we may apply certain rules to those statements in order to derive brand new statements that are guaranteed to be true as well. These new statements are logical deductive inferences. Here is an example of first order predicate logic syntax:

Given: $\forall(x)$ kid $(x) \Rightarrow$ loves-candy (x)

And: $\exists(y)$ named $(y, \text{Jimmy}) \wedge$ kid (y)

Then: loves-candy (Jimmy)

The syntax translates to plain English in the following manner. **Given:** *For all x, if x is a kid, then x loves candy*; **And:** *There exists some y such that y is named Jimmy and y is a kid*; **Then:** *Jimmy loves candy*.

Once we have a list of facts and relationships properly expressed in the syntax of first order predicate logic, a computer can easily chase through the possible combinations of relationships that lead to inferences. Derivations like this require the application of some odd rules known as *skolemization* and *unification*. So, we won't go into any more detail on first order predicate logic. But there are some obvious associated shortcomings that we need to discuss.

While first order predicate logic is capable of automating the creative process of drawing inferences under certain circumstances, it is terribly insufficient for capturing all the knowledge that we humans carry around in our brains. One severe limitation relates to an inability to use the relationships expressed by first order predicate logic as objects themselves in higher order relationships. For that, we need a higher-order predicate logic. If it is true that the human cortical organization is indeed based around nested relational hierarchies, then first order predicate logic allows us to model only a single level of what the cortical hierarchy encodes. Just as the brain has many levels of hierarchical pattern recognition, so must any language of artificial intelligence be able to express patterns of patterns of patterns, and so on.

Another limitation of first order predicate logic relates to the fact that we don't know anything with complete certainty. Everything we humans believe carries with it some probability of validity. But, first order predicate logic has no means for expressing the associated level of validity for any given statement. There have been some attempts at dealing with probabilistic validation by a class of systems typically referred to as 'fuzzy logic'. But they also suffer from some intractable limitations.

Yet another limitation relates to the inability to express dynamic relationships. So, even if we were to extend first order predicate logic to Nth-order fuzzy logic, we would also need to add an ability for sequentially linking various relationships into dynamic logical entities.

Despite these limitations, computer scientists have made significant progress toward programming machines to think creatively in highly specific domains. For example, by searching all possible variable bindings in abstract logical expressions, computers have indeed come up with brand new and useful theorems and proofs. In fact, in 1996 a computer found a proof for Boolean algebra using axioms that had been suspected of being true for sixty years. Whereas humans were unable to prove the axioms during that time, the computer did it after eight days of computation (McCune 1997).

The Logic of Implication

Let us now turn our attention to a particular sort of deductive inference. We'll consider the logic of *implication*. This is of interest to us because implications are somewhat analogous to causal memes. Perhaps we can use the formal logic of implication to implement the sort of creativity, based on causality, exhibited by humans. Here is how implication works: The statement '*A implies B*' simply declares that if A is true then B is necessarily also true. For instance, being a man implies having a Y-chromosome. So, if it is true that a particular individual is a man, then it must also be true that he has a Y-chromosome.

Now, it should not be difficult to see that if *A implies B* and *B implies C* then we can easily infer that *A implies C*. This sort of deductive inference can be attained through a logical inference known as *modus ponens* (applied recursively). Such inferences can be chained together whenever the consequent (output) of one implication matches the antecedent (input) of another. Stringing together implications, just as I have analogized the stringing together of causal memes, is like arranging strings of dominos so that adjoining ends have identical dot patterns. Computers can easily be programmed to methodically try every possible combination, thereby finding every possible inference of implication. Given a list of implication statements expressing knowledge about the real world, we might expect there to be a great many possible inferences.

Implications are similar to causal memes, but not strictly identical or as powerful. They capture a static characteristic of reality. Implications are conditionals: If *this* is true, then *that* is true. Causal memes are also conditionals, but they include a component of time: If this is true, then that *will be* true. But implications may also imply future states of the world. The real value of causal memes over logical implications relates to the sorts of arguments on which they operate. Causal memes have antecedents and consequents (inputs and outputs) that can be very complicated expressions of nested relationships involving the smallest of details across many different dimensions. Those dimensions, in addition to the three dimensions of spatial size and shape, include such attributes as color, texture, hardness, weight, flexibility, smell, taste, sound, temperature, and many more.

The Nth-order relationships that comprise the inputs and outputs of causal memes define points in some high-dimensional space. They automatically encode things in terms of all the different attributes measurable by our various senses. There is no way to extend logical implications to operate on such complex arguments without fully embracing the hierarchical structure of the human cortex. The more we understand the requirements for computational representation of knowledge, the more we can appreciate the power of cortical hierarchies. If our goal is to invent human-like computational intelligence, there is no getting around the need for powerful expressions built out of Nth-order relationships. So, let us assume that the only way to achieve human-like intelligence is to build a human-like structure of knowledge, which entails something similar to the relational hierarchies of the human cortex. But here lies a problem.

Circuit designers tend to build electronic systems out of discrete packages, called integrated circuit chips. These chips, while extremely complex inside, are limited to a handful of external connections. Unfortunately, there is no obvious way to carve out a chunk of cortical functionality without there being millions of required external connections to that chunk. And the construction of a single, monolithic system is also problematic due to the amount of functional defects that are likely to occur within such a large system. Luckily, the inherent redundancy of the architecture may allow us to build monolithic systems with an ability to disable or ignore any defective individual neuronal correlates.

There is no doubt in my mind that exploration of cortical functionality will first occur through software simulation. Given the huge number of neurons and the associated computational power required for simulating the cortex, the initial simulations will be extremely slow. But once we understand how the system works, through a Lamarckian style of intelligently directed trial and evaluation, we can then design specialized hardware to execute the simulation much more quickly. Jeff Hawkins has formed a new company called *Numenta* that is already engaged in such an endeavor.

We will certainly simulate the architectural workings of the human brain when we have the computational power to do so. And that required amount of computational power will probably exist within the next thirty years, or less, perhaps much less. Once we are able to simulate the architectural characteristics of a human brain, we will likely feed it sensory data as if it were a real brain in a real environment. It is the logical next big step in the evolution of technology. As I previously hinted, these facts are critical to an upcoming discussion of cosmology. Please keep them in mind.

Intuitive Reasoning as Constraint Satisfaction

Intuitive reasoning may be viewed as an evolutionary process of constraint satisfaction within the context of conceptual hierarchies. I speculate that, when a mind is presented with some unusual data, it inductively tries to make sense of that data by determining whatever interpretations minimize the extraneous assumptions. In essence, the mind embraces the principle of Ockham's razor in an effort to find the leanest possible interpretation of any unusual situation. It effectively constrains interpretations on the basis of consistency with the beliefs it already holds as valid. By striving for consistency among its beliefs, the mind tries to ensure the validity of its beliefs.

The brain automatically searches through various complementary combinations of the many learned relationships about which it knows in order to interpret incoming sensory data. It tries to minimize the difference between current sensory patterns and various combinations of known patterns. The relationships we know and hold as valid act as interacting opportunities and constraints on the problem to be solved. They interact in some N-dimensional space, producing a sort of 'landscape' within that space, and the problem amounts to finding the global minimum of the landscape – the point at which the difference between what we perceive and some complementary combination of things we know is minimized. A brain may probe the landscape at any coordinate, but it cannot 'see' the entire landscape without probing every possible coordinate, which can be an impossible task, if there is a huge number of possible combinations.

Finding the global optimum is a common problem in artificial intelligence. Indeed, for any problem, finding the most intelligent solution amounts to finding the optimal solution. So, the nature of all problem solving is to find the global minimum (or maximum) of some function that describes the problem. Yet, most such functions produce 'landscapes' that are simply much too large to thoroughly search. Luckily, there is a shortcut way of searching for the global minimum of a problem landscape through a process known as 'simulated annealing'. The process works well in most but not all cases. The term 'annealing' refers to a practical real-world method of getting the atoms in a piece of metal to pack themselves in their lowest energy state – thereby achieving optimum stress-free ductility – by first heating the piece of metal and then allowing it to cool very slowly.

The process of *simulated* annealing can be roughly visualized as follows: Put a 'Mexican jumping bean' on the mathematical 'landscape' of a problem to be solved. The bean hops around, randomly sampling various coordinates and 'sliding' down 'slopes' into various basins of *local* minima. The 'sliding' happens by a continuous process of sampling immediately surrounding spots and moving in the direction of the lowest, until all surrounding spots are higher than the current spot. The height of the bean's jumping depends on its energy level. Theory tells us that, by reducing the energy of the jumping bean very gradually, we can guarantee it will eventually find, and stay in, the valley of the *global* minimum.

However, the amount of time implied by a very gradual reduction of energy could be extremely long. The more gradually we reduce the energy in the jumping bean, the more likely we are to find optimality – the absolute lowest point in the landscape.

When we consider this heuristic technique in regard to the brain, we see that the associated pseudo-random noise in the firings of nonlinear neural processes may correlate with the random explorations of the problem landscape through a process that analogizes to simulated annealing. As neural firings 'hop' around loops of connections, the activity explores the connection 'landscape'. As the fatiguing effects of neurons cause a gradual long-term lessening of neural activity, the system slowly looses energy and will, according to theory, eventually settle into a global minimum. Of course, real brains can't spend an infinite amount of time on a problem, so there is a practical trade-off between the speed and the fitness of a result.

Let us try to get a little more specific. Given some unusual situation, the brain is tasked with trying to understand it. This amounts to a process of probing the many complementary combinations of the millions of patterns held by the brain in order to find the one that best fits the situation being sensed. Now, consider the following very unusual example.

Suppose you were to see an image of an elephant in a jail cell. The elephant is partially obscured by the bars in front of it, yet the mind easily separates the bars from the elephant and fills in the partially obscured view. Let's review how it works: First-responder circuits detect partial elements of the elephant, enough to excite the set of neurons at a high level in the cortical hierarchy that symbolically represents the visual aspect of an elephant. The corresponding set of high-level neurons sends neural activity back down the hierarchy to all the lower-level neurons that have tended to correspond to the recognition of elephants in the past. The pattern of activity describes what a complete elephant should look like.

This style of feedback allows lower-level neurons to compare what they currently see with what they would expect to see when looking at a typical, unincarcerated elephant. Those expectations amount to predictions based on the history of previous exposures to unobscured elephants. And so, consciousness *sees* the partially obscured image, but *imagines* that there is a complete elephant there. The vertical bars strongly violate the downward-propagating expectations, and are therefore attended to as unusual. They excite another hierarchy corresponding to the recognition of a jail cell.

As we consider ascending levels of the cortical hierarchy, we find that higher and higher levels are likely to be more and more divorced from sensed reality, but more and more corresponding with actual reality. The low-level sensory data paints the vertical stripes of the bars right on the elephant, as if it were a new species of fat zebra. The understanding of actual reality requires the mental separation of the two visually overlapping symbolic entities – the elephant and the jail cell. The neural activity at higher cortical levels does indeed separate the bars from the elephant. At those higher levels, the neural sites corresponding to both a jail cell and an elephant are actively firing. And, at some cortical level, the brain makes preliminary sense of the data by separating the bars from the elephant.

Let us presume that the process allowing us to make sense of unusual data is one of trying out many combinations of things about which we know. Once again, we look to a meiotic style of evolution as a means for achieving intelligence. In the specific example, it shouldn't take too long before settling into the conceptualization of two separate entities – an elephant and a jail cell – because the shapes involved are rather unique. But what about the even higher cortical levels? How can the brain make any sense at all of an elephant behind bars?

The answer is the same at all levels of the cortical hierarchy. Making sense of incoming signals at any given level depends on some neurons being sensitive to the pattern exhibited by currently active neurons at the lower-level. In our unusual example, the cortex merely needs to have witnessed such a simultaneous occurrence of an elephant and a jail cell in some previous context. There has to have been a previously formed relationship that links elephants and jail cells. The brain can then infer from the incoming data the possibility that it is witnessing the same higher-level symbolic entity to which it had previously been exposed. And, when it does so, it sends downward flowing activity to all the other associated hierarchies that were active during the previous experience.

With respect to the elephant behind bars, my mind has two possible high-level symbolic entities that fit this simultaneous superposition of an elephant and a jail cell. Perhaps the jail cell is actually a cage on a train car that is used by a circus to transport the elephant from town to town. I might not be able to make this connection had I not seen the movie *The Greatest Show On Earth*. Recognize the similarity between a large cage and a jail cell. Their conceptual hierarchies would certainly share many abstract nodes, and would, therefore, be logical candidates for exploration in the process of reconciling the image.

Or, perhaps the image of an elephant in jail makes sense if I am watching an episode of *The Andy Griffith Show*. I wouldn't put it past deputy Barney Fife to lock up an elephant for jay-walking. Again, I would not have been able to make the connection without having seen many episodes of the show. But, having seen *The Greatest Show On Earth* and many episodes of *The Andy Griffith Show*, including one episode in which Barney does lock up an animal, my mind has at least two abstract, high-level hierarchies that will resonate in response to an elephant behind bars. Additional information allows my brain to choose between them.

Suppose I were to walk into a room and see on a television the image of an elephant behind bars. If the image is in black and white, my mind will automatically collapse its possible interpretations down to the *Andy Griffith Show*. But, if the image is in color, then I'll assume it is the movie *The Greatest Show On Earth*. My reactions and subsequent thoughts depend completely on my past experiences, and how their neural encodings are able to resolve ambiguities in what I am currently sensing.

One's conscious perception of reality at any moment reflects the single best fit, regarding previously acquired sets of patterns, to patterns being sensed at the lowest levels. The brain always seeks to optimize the satisfaction of its constraints.

If you have no experience with these shows, then perhaps you have your own relevant experiences on which to draw. Perhaps the silly idea of an elephant in a jail cell conjures up a completely different sort of past experience for you. If not, then perhaps you can better appreciate the fact that current experiences are only interpretable in terms of previous experiences. And, if there are no relevant previous experiences, then there can be no understanding of current experiences. If nothing comes to mind for you, then perhaps you have never previously thought about an elephant in a jail cell, and therefore, you have no means for understanding how such a bizarre situation could have come about.

The constraint satisfaction performed by neural processes, and the consequent collapsing of possible interpretations, is even more evident when we consider the many interpretations that are available to situations that are only imagined. Merely reading the word 'elephant' can cause its corresponding symbolic neurons to fire, and those firings send downward excitations to all visual patterns that would be likely to fire if actually looking at an elephant. When imagining an elephant, the brain has no excitation from first-responders, yet the image of a prototypical elephant is captured by resonating circuits behind the 'visual curtain'. An imagined elephant is 'sort of' in the conscious experience, even though it isn't in the visual field.

The mentally conjured prototypical image is not like a picture of an elephant, but rather, like a bunch of movies each taken from a different angle and each capturing a different elephant activity. The subconscious mind *feels* them all simultaneously: an elephant walking, standing, wagging its thin tail, and swaying its thick trunk – they are all there at the same time. With additional information, the multitude of image sequences collapses down to a single view. When confronted with the mental suggestion of an elephant in a jail cell, the set of mental views collapses in my mind to a side view of an elephant standing still behind the bars. After all, my mind is constrained by the simple belief that the elephant can't be walking if it is in a jail cell.

After bouncing around between all possible interpretations, like a jumping bean on a hilly landscape, firing patterns gradually loose their energy and the brain thereby settles on whatever interpretation makes the most sense – whatever reconciles best with all other abstract beliefs.

As I further consider why the elephant might be in jail, the mental perspective changes in my mind to a frontal view of an elephant peering through the bars, as if it were saying "I've been framed." My complete mental set of views has collapsed to just a subset of the possible views I hold for a typical elephant. And, my brain may automatically apply abstract mutations – in this case, anthropomorphizing the elephant – to see if it helps make sense of the situation. With more information, my view will collapse even further. Consciousness is the collapsed view. It is the single state of re-responder neural activity that is most consistent with incoming sensory data.

Just as creativity results from the brain's ability to evolutionarily probe various possible combinations of possible scenarios, so must inductive constraint satisfaction result from similar evolutionary probing. But the goal for constraint satisfaction is not to maximize happiness, but to minimize contradiction. As more and more information becomes available, possible interpretations collapse down to whichever one minimizes both extraneous assumptions and contradictions with held beliefs.

The collapsing of uncertainty with regard to various interpretations of sensory data reminds me of the collapsing wave function in quantum dynamics. At the quantum level, Schrödinger's wave function predicts many possible states, corresponding to Heisenberg's uncertainty principle. The Copenhagen interpretation of quantum dynamics declares that the uncertainty collapses down to a single state only as a result of observer participancy. That is, nature doesn't decide among the many possible universes that uncertainty creates until it absolutely has to – until observer participancy constrains it. And, likewise, my mental image of reality doesn't resolve among the many possible interpretations until it absolutely must – until further information constrains it to a single view. I'll soon show surprising evidence that the uncertainty of nature may indeed be intimately connected with the uncertainty of mental processes.

Let us now turn our attention to the final chasm between computers and truly useful intelligence. It is the inability of computers to properly process natural human language. This is the area in which the ideas of this book are likely to have their most practical and dramatic impact. Although, please don't expect me to solve the problem here.

Natural Language Processing

Computers have gotten extremely intelligent in specific domains. Indeed, they are now very good at playing the sophisticated game of chess. A landmark event occurred in 1997 when a computer defeated the world chess champion Gary Kasparov. Despite such an amazing achievement, most of us don't think of computers as being intelligent for the simple reason that they can't yet speak our language. Computers have trouble communicating with humans because our natural languages are full of exceptions, ever changing slang expressions, jargon and idioms that defy generalization by rules.

Humans literally spend a lifetime learning human language, and computers have not yet even existed for a full human lifetime. So it should not surprise us to discover that computer scientists are on the wrong track when it comes to processing language. Since they successfully model all the laws of nature by strict mathematical rules, they seem to similarly believe they can model language that way. But, because language evolves memetically, we shouldn't expect it to be completely rule-based. We should expect it to be highly complex, intricate and full of idioms, slang, jargon, repeated malapropisms and colloquialisms. Biological evolution produces animals that are highly intricate and full of special purpose devices. Indeed, there is no simple set of rules for describing the human body, as evidenced by the fact that very few humans are exactly identical. We shouldn't expect the evolution of complex memetic systems, such as language, to be significantly different. Yet computer scientists and linguists continue searching for the rules of language.

It is my well-considered opinion that we won't make notable progress on computerized natural language processing until we have first developed techniques that *embrace* the diverse range of anomalies within a typical human language. Instead of searching for rules of language, we should build computers to remember all patterns of word usage to which they are exposed and use the most common of those many diverse relationships and patterns as templates for constructing new sentences. Such a brute force style of approach necessarily involves the analysis of many commonly constructed sentences and the use of a huge database of interconnected general knowledge – memetic knowledge – similar to that which we humans carry around in our brains.

The quickest path to achieving a system capable of understanding natural language will likely be by designing computer architectures around the same sorts of relational hierarchies as are apparently encoded by the human cortex. But how, then, can we map English sentences into that style of architecture? It is actually quite straightforward, although a bit difficult to describe.

Let us consider the design of a system that can read books directly from computer files full of text. The system will scan a file, one sentence at a time, organizing its hierarchical structure on the basis of all the relationships between characters and, at a higher level, between words. Such a system effectively by-passes the first few levels of the human visual system, because it doesn't need to deal with the representations of elements as small as individual photoreceptors, edges, and so on. At some level in the human cortical hierarchy, text letters are represented by specific nodes, and that is the level at which our computed hierarchy begins. The organization of all hierarchical levels above text characters will be similar for both the cortical system and the computed system.

To educate the artificial system, we expose it to millions of text sentences, one at a time, from lots of text files representing books. The first level of artificial cortex can be thought of as a flat array of nodes with its length dimension indexed by the character position along the current sentence, and its width defined by all the possible characters. You should be visualizing a flat array of nodes with the 'current sentence' buffer beside it, running along its length. At each character position along the sentence, one node across the width of the array will be active, corresponding to whatever the character is at that position. The entire sentence is thus represented by the active nodes in the first level array.

Higher levels are represented by arrays of cells (perhaps much more densely packed) overlaying the first-level array. Each array simulates a level of cortical functionality. So, each cell in any of the higher-level arrays must be capable of monitoring activity within some region of cells, centered directly below it, in the just-lower layer. For example, a cell in the second layer might monitor, as if by dendrite-like sensory 'fingers', a ten-by-ten array of cells centered just below it. For every cell in the second layer, the monitoring 'fingers' are weighted differently for each lower-level cell being monitored. In similar manner to a synaptic neural connection, the weighting factor for each of the monitoring 'fingers' of a particular cell, must grow stronger with simultaneous 'firings' between the cell itself and the lower-level cell that it monitors. Additionally, for layers above the first, when a cell 'fires', that event inhibits the firing of other cells in its own layer within some region that we have referred to as a region of orthogonality.

We may logically suspect that, at the second level, simultaneous activity from below will automatically organize nodes to recognize common combinations of characters, such as: 'th', 'ing', 're', 'tion', and so on. Of course, we can't pretend to know exactly how the divvying-up will go until we program a simulation, set the various parameters, and feed it a large corpus of text. But there are some inevitable problems that we can quickly identify. For example, a different node corresponding to the same common combination of letters must eventually emerge at almost every spot along the length of the sentence, because each character combination can occur at nearly any spot in a sentence. And further, there must be a node for each word at every possible location along the length of the sentence. This redundancy is necessary because any particular word can occur at nearly any spot in the sentence.

Perhaps we can alleviate the need for much of the redundancy by having the current sentence scroll through its buffer, as if it were a 'walking' marquee sign. My goal here is not to design a working system, but rather to illustrate some important high-level concepts and problems. Thus, I won't engage in the tedium of developing details of the model any further. You get the idea.

Words will be represented at levels above common character combinations, and common phrases will be represented at higher levels, above words. There are quite a few semantically decomposable phrases in the English language. For example, the phrases 'kick the bucket' or 'down the drain' tend to lose their abstract meanings when their words are not juxtaposed. At still higher levels, complete sentences will be represented by nodes having links to the words and phrases that make them up. For instance, there is probably a node in your brain representing the sentence: "Ask not what your country can do for you; ask what you can do for your country." And, perhaps another node for: "I did not have sexual relations with that woman ..."

Suppose that every sentence you have ever heard, read, spoken or thought is represented by a node in your brain. Those sentence nodes are also likely to have links to other nodes representing the authors of those sentences. So, if you read a certain sentence that you have previously heard, the author of that sentence can automatically spring to mind. Did that happen with you in the two examples of famous phrases I just cited?

I certainly can't remember every sentence uttered by my first-grade teacher, but that doesn't mean the corresponding nodes no longer exist in my brain. I'm sure she taught me a great many concepts that I still know. The corresponding abstract nodes have been reinforced over the years by many other teachers and situations, thus diluting the connections to the node that represents her. But she probably planted the seeds of many nodes that still exist in my brain and that form my general foundation of scholastic knowledge. She merely gave names to others that had automatically formed, such as certain aspects of language with which I had intuitively become familiar. Consider an example of such intuitive learning.

It turns out that all the words of a particular 'part of speech' can become automatically linked into a very broad category. For instance, all words representing people, places and things become linked to a common class we call *nouns*. How do they automatically get associated? The answer lies in the fact that all nouns tend to occur after the very same common words: the articles 'a', 'an' and 'the'. If a newborn brain hears the phrase 'the wagon', then the next time it hears the word 'the' it will predict the word 'wagon'. If the same brain then hears the phrase 'the truck', it will have predicted the word 'wagon' but it will have heard the word 'truck'. The corresponding nodes for both words, 'wagon' and 'truck', will be active and will thus become linked at a higher level. All nouns can become similarly linked into an overarching category after some time. Similar language cues probably exist for classifying all parts of speech.

As words become classified as nouns, verbs, adjectives and so on, the brain is then capable of collecting sentence patterns in terms of those various parts of speech. Those syntactic sentence patterns can be used to properly compose brand new sentences. Further, various types of phrases and sentence fragments may combine in certain patterns of patterns, as indicated by one's history of exposure to them. The structure of a language becomes represented completely in terms of the many commonly occurring abstract hierarchical relationships between the various primitives of the language.

Notice that language syntax is not defined by rules, but rather, by common patterns of usage. There certainly are some generalized patterns in the ways various parts of speech are typically linked together. And those patterns will likely be replicated over and over again in the production of brand new sentences, simply because our minds are built to mimic patterns at all levels. Some of those patterns of syntax may indeed be expressed as rules. But specific violations of those patterns of grammar, such as can occur in slang terms, jargon, malapropisms and errors of speech, can also be mimicked. And when those exceptions are heavily mimicked by others, they achieve as much grammatical validity as any other pattern of grammar. In such a manner, grammar evolves.

Once we are able to computationally map English sentences of text into overlapping relational hierarchies, then we will be able to automatically identify complementary interfaces to causal concepts by the amount of overlap in their hierarchies. And we can then combine those concepts, to produce new, bigger and potentially better concepts. Recall that useful memetic mutations are more likely to result from various sorts of combinations of complementary memes than from random errors. By meiotically combining causal memes and simulating their causal effects, computers can become as generally intelligent and creative as humans. I suspect a computational system that scans and compiles books into relational hierarchies, like the one we just discussed, will be built within ten or twenty years. It will be extremely slow at first, but it will mark the beginning of human-like artificial intelligence.

Anyone who has studied the computer science field of natural language processing knows that it is extremely complicated. I have only touched on several of the issues. At bottom, I simply suggest that every one of the difficult language issues is handled by the human brain on the basis of abstract patterns it automatically collects. We can't possibly understand the details of how the brain does some of the things it does without building and examining a machine that similarly compiles patterns of things it experiences. Only then can we appreciate the sorts of abstract constructs available for algorithmically computing intelligence.

There is some circumstantial evidence that we are on the right track with hierarchical relationships, coming from the area of artificial intelligence concerned with speech recognition. The intent of speech recognition research is to program a computer to analyze an electronic signal coming from a microphone into which someone speaks. By using the previously described mathematical tool, the Fourier Transform, a computer is able to roughly detect various syllables and consonants, which are then combined into words.

The big problem in speech recognition research relates to the automatic detection of when one word ends and another begins. When we humans speak, we often run our words together. Therefore, computers have difficulty differentiating between the spoken versions of, for example, 'categorize' and 'cat egg or eyes'. We humans are able to tell the difference based on semantics. When we hear the word 'categorize', we know it is not 'cat egg or eyes' because there is no appropriate semantic interpretation of 'cat egg or eyes'. It just doesn't make any sense. So, we correctly assume the speaker said 'categorize'.

The computation of semantic interpretation has been out of reach for programmers in the past. So, they have gotten around the problem by using statistical analysis. They build statistical tables, known as *Markov models*, that essentially express the likelihood of one word following another. Here's how they do it:

Imagine a huge 2-dimensional array of numbers initially set to zero. The rows of the array are indexed by all the words in the English language, ordered alphabetically. So, the first row would correspond to 'aardvark' and the last to 'zoom'. The columns are identically indexed. Now, the computer scans a large corpus of text, and, for every word combination in the text, increments the cell corresponding to that combination, where the row is indexed by the first word in the combination and the column is indexed by the second. The final table holds information regarding the respective probabilities for any word being followed by any other word. We can use the table to assess the relative likelihood of any string of words in a sentence. We would quickly discover that the string of words 'cat egg or eyes' is extremely unlikely.

Markov models have been so successful as to cause some researchers to speculate that the human brain is essentially just a statistical device. Indeed, when we consider how Hebbian learning works, we see that the strengthening of a connection between the nodes representing two juxtaposed words in a sentence is just like the incrementing of a cell in the previously described statistical table. In fact, such a statistical table approximately represents a single level of the relational hierarchy that we presume to exist in the human cortex (at the level of words). So, we shouldn't be surprised that Markov models have been so useful in speech recognition. But a single level just isn't very interesting. What we need are Nth-order statistical tables. But, what, exactly, would that entail?

There are such things as Nth-order Markov models, but I suspect they won't do the trick. A second-order Markov model is one that considers the probabilities of various words occurring as a function of the previous *two* words. Such a model would be even better at predicting the rarity of such three word combinations as 'cat egg or' and 'egg or eyes'. A third-order model considers the previous *three* words, and so on. An Nth-order Markov model is clearly not the same thing as an Nth-order relational hierarchy. It is not nearly as expressive. Nor does it facilitate a metric of *similarity* between abstract concepts, as can be discovered by the amount of overlap between two conceptual hierarchies. Unfortunately, the expressive power of Nth-order hierarchies comes with a steep cost of enormous storage and processing power. Yet, with every year of advancing technology we get closer to being able to afford those costs.

We will have reached the age of truly intelligent machines when computers are able to digest the text of a book on any subject, and then respond intelligently to questions having answers that are only *implicitly* stated within the book. Such a feat will require the ability to draw logical deductive inferences, as previously discussed. But we are finally beginning to understand just how that might be computationally possible.

Chapter 10. **Universal Darwinism**

When Charles Darwin developed his theory of evolution, it was destined to stand the world on its ear. That simple yet powerful idea completely altered the thinking of many intelligent minds and sparked righteous indignation in other unyielding minds. Since then, scientists have discovered mountains of evidence strongly supporting Darwin's theory and have even learned to read and manipulate the genetic patterns that exist in *every* living organism.

We rightly adorn the legacy of Darwin with credit laureate, as if he gave us something that we would otherwise never have had. But a profound consequence of his discovery, as it applies to the memetic replicators of thought, now makes clear that the concept of evolution was destined to emerge at that time, if not from Darwin's mind, then from the mind of, say, Alfred Russel Wallace. Yes, Darwin deserves credit for studying various aspects of life and collecting the many complementary memes that would automatically combine in his mind to form his famous theory. But we must not lose sight of the fact that the event was neither spiritually nor even volitionally inspired. It was statistically destined to occur. It stood directly in the path of evolution's accelerating arrow.

The evidence of science overwhelmingly supports a single, simple, foundational idea, best described as *Universal Darwinism*. It is summarized by the one simple sentence I have already recited several times: There is no source of creativity anywhere in the universe other than the process of evolution. The Darwinian process, applied to various sorts of patterns, accounts for everything related to life and intelligence, including even immune system antibodies, the wiring of neurons, the firing of neurons, culture, language, religion, beliefs, technology, art, and even the natural emergence of morality.

A Summary of the Memetic Perspective

If a human mind is as deterministically bound to the trajectory of physics as science tells us it is, there can be no logical way of looking at intelligence other than as a statistically determined evolutionary process operating on patterns of thought. Yet, very few books on intelligence give any nod at all to the concept of memes. And, even those that do mention memes tend to limit the attention to only a few pages, or less.

Memes have been defined here as patterns that serve as templates for their own replication or translation. Using that definition, patterns of behavior become memes whenever they are viewed and mimicked by others, and patterns of thought become memes whenever they are later reconsidered or communicated to others by language. Both patterns of behavior and patterns of language are automatically translatable into corresponding patterns of thought. And, once they are translated to patterns of thought, they are able to evolve by way of automatic cognitive mechanisms.

Patterns of thought always exist as patterns of neural activity that can recur over and over again, sometimes combining and recombining in various ways, and sometimes producing imagined plans that are expected to yield significant benefit. By broadening the definition of memes, we are able to unify culture, technology and human intelligence into a single phenomenon, all operating by way of the same fundamental mechanism – patterns of neural activity, evolving within a single brain and, through many sorts of translations, across many brains.

Hard-wired emotions give replicating patterns of thought access to manipulation of the human machinery, and the thought patterns that are best able to stimulate the triggers of those automatic emotions are the ones that find themselves hopping from brain to brain, deterministically causing those brains to direct energy toward the perpetuation of the very same patterns of thought. Memes tug on our heart strings, push our buttons of anger, flirt with our lusts, terrorize our fears and deliciously tempt our hungers. We don't choose to feel our emotions, they just automatically happen under certain circumstances. The automatic character of our genetically defined human emotions further supports the memetic perspective. Indeed, patterns of thought can evolve so as to prey on the hard-wired emotions of humans, thereby fulfilling their very own agendas of perpetuation.

All our human emotional feelings of consciousness, including hunger, lust, pride, shame, jealousy, gratitude and love, are genetically inspired to occur under certain conditions, and serve as absolute anchors for our elaborate plans. They give us all our wants and desires, and they thereby form the base motivations for everything we do. Genes define the things we want, and memes define the ways we go about getting the things we want. Our conscious feelings are the mechanisms by which our genes ensure that our human wills are consistent with their perpetuation interests – the interests of our genes.

Our brains automatically try to maximize, over the long term, pleasurable feelings, and to minimize, over the long term, unpleasant feelings. Our brains assess ideas as being *good* when they promise to bring pleasure. And, even when the payoff is promised to occur after death, to a brain that believes in life after death it is still a good idea. Bad ideas are those that are expected to result in the possibility of pain or discomfort.

Genes had a strong evolutionary motive for giving offspring the ability to mimic the behaviors of their parents. Such an ability to hand-down behaviors caused those behaviors to quickly evolve, and to become adaptive. It was a faster and better style of evolution. But there would have been no evolutionary reason to allow mimicry to be a matter of free will. Natural selection would have quickly eliminated any genes whose hosts had the freedom to choose whether or not to mimic parental behaviors on a whimsical sort of will. Evolutionary success requires the choice – of whether or not to mimic parental behaviors – to be made always in the interest of perpetuating the genes that enable the mimicry. The choice to mimic has to be rational. It has to be automatic. And it has to benefit the genes.

Simple mimicry naturally evolves toward intelligent mimicry through which minds can mentally simulate and thereby evaluate the effects of combining various actions, many of which have been previously acquired through simple mimicry. An internal evolution of thought allows many different combinations of actions to be contemplated, and the best to be selected for performance. Minds then become capable of creative thought, leading to proactive behaviors, which are evolutionarily preferred over reactive behaviors.

Prediction is the key to intelligent mimicry, and it relies completely on an ability to accurately simulate the world, mentally. Accurately predicting the future under various scenarios is the essence of what intelligence was designed by evolution to do. We humans, more so than any other species, make our decisions by mentally simulating various options in order to identify which paths will likely lead to the most future happiness.

Simulation is, in essence, a mind's ability to internally mimic repetitive aspects of the world. We humans are able to simulate the world and thereby predict its future states only because we live in a very deterministic world and because we have very deterministic minds. A deterministic mind is the only style of mind that can simulate the deterministic reality of the universe in which we live. It is also the only style of mind that can exercise rational judgment and exhibit reliable behavior. This is a hugely important philosophical revelation that allows us to see the incredible value of deterministic thinking. It is to be embraced, not rejected or feared.

It now appears as though the ability to mentally simulate the environment can spontaneously result from the very simple algorithm of gathering statistics through Hebbian learning. The human brain compiles statistics about the world during every wakeful moment of its life experiences. Those statistics ultimately manifest themselves as our beliefs in causality.

Most of our beliefs are automatically acquired after birth through experience. But a few of them – our beliefs in our own conscious feelings under certain conditions – are innately provided to us by our genes. We automatically, and sometimes subconsciously, construct our plans by stringing together causal beliefs. At the ends of all those strings are innate beliefs in feelings of pleasure. Beliefs are the currency of thought. Our brains deal in nothing but beliefs. So, any genetically supplied motivations must come in the form of beliefs. Indeed they do.

We rationally construct our plans so that they will statistically yield benefit to ourselves, which can only come in the form of our beliefs in our own good feelings of pleasure and happiness. We believe the benefit of pleasure accrues to ourselves as individuals, but in actuality, the benefit statistically accrues to our genes. Such is the mechanism by which our genes force us to act in their interests, all the while, causing us to believe we act in our own interests.

While many people find this duplicity to be depressing, there is a brighter side to things. We may take comfort in the way nature has caused our genes to define our morality. As true as the old adage is that 'nature abhors a vacuum', so is it also true that nature loves cooperation. Such is the basis for our moral aspirations. All our moral precepts have evolved so as to facilitate cooperation.

Through the memetic replication of behaviors, involving especially the learning of language, individual efforts can become coordinated into various forms of cooperation that can yield benefit to all involved. Cultural societies thereby emerge. Societal memes of culture are able to evolve in various directions, and the direction that the evolution of those memes ultimately takes is determined by whatever systems best facilitate synergistic cooperation.

From a scientific perspective, it is the facilitation of life that defines the moral good. And the facilitation of life is best accomplished through the discovery of better memetic behaviors for facilitating cooperation. Early on, cultural practices of reciprocation teamed up with the genetically supplied emotions of gratitude and revenge so as to encourage people to cooperate with each other in the spirit of Tit-for-Tat. Now, governmental memes (laws) ensure ever greater cooperation.

Our modern environment encourages us to collectively conceive of ever deeper technological memes that ultimately serve to direct energy toward the perpetuation of all our defining replicators. A democracy encourages efficient cooperation toward the development of technology more than, say, a dictatorship does. For that reason, democracies are ultimately destined to prevail over dictatorships. And, likewise, a capitalistic society encourages technology memes to evolve more rapidly than, say, a communistic society does; and so, capitalism is destined to flourish, until something more conducive to memetic evolution comes along.

Intrinsic value is defined by any sort of cooperative contribution to the evolutionary development of aggregate life, which, at the human level, relies on *valid* memetic beliefs and beneficial memes of technology. If there is to be any rebellion against the tyranny of selfish replicators, it is to be directed toward popular memes that are *invalid*. For, it is those memes, and those memes alone, that can damage the health of our cooperating society and can cause the elimination of cooperative replicators in general.

There should be no reason to rebel against our selfish genes, unless we are prepared to forsake our children. And, there should be no reason to rebel against the *valid* memes of technology, for, they hold our collective destiny, and they are perfectly willing to share it with us in advance if we are willing to learn their language – the mathematical and logical languages of the hard sciences.

A Higher Calling

By now you should have experienced a flip in your perspective on genes versus memes. Most people, on discovering the idea of memes, tend to think of them as being in some sense secondary to the primary replicators, the genes. The flip in perspective that I'm hoping you would have experienced by now places memes as the ultimate replicators. They subsume genes. Patterns are nothing more than just patterns, independent of the substrate in which they exist. So, how should this affect our view of ourselves? Are we more defined by our genes or by our memes? Which is more important, our biological bodies or our ideas? If it isn't already clear to you, let me illustrate with a thought experiment.

Suppose there existed a pill, which, when taken daily like a vitamin, would allow a pregnant woman to increase the health, intelligence, physical strength, and attractiveness of her child. In my hypothetical scenario the effects would be minor, but they would be measurably better. Should she take the pill? Since many women take vitamins for the very same purpose with absolutely no understanding of how they work, I am assuming the answer, by many, would be 'yes'.

Further suppose that the babies born to women who take our hypothetical pill are physically different than normal babies. Well, I believe that many people would expect a better baby to be different, physically. Now, suppose it were later discovered that the brains of the special babies have some rare substance in them that allows them to be smarter brains. And, suppose the pill had been used by many women with only successful results. Still okay, right? Suppose that the use of the pill were to become so common that, just to have a baby of average IQ, a mother would need to take the pill. Then, a mother had better take it.

Now, suppose it were discovered that the strange substance in the baby's brain was actually composed of silicon circuits that operated electrically just as computers do. And, suppose that successive generations of births, augmented by the pill, were to result in smarter and better babies having higher concentrations of silicon in their brains. Should the mother of each generation not be proud of her better baby? Eventually, in this hypothetical scenario, humans would evolve to become silicon-brained humans, much smarter, yet still equipped with all the moral emotions of love, honor, compassion, appreciation of beauty and so on. We would have given birth to a new and better branch of life. Now, suppose that scientists were able to develop technology for building silicon-brained humans from scratch, bypassing the painful process of birth. Who could argue that it wouldn't be a better system, or do we prefer to see women suffer during child birth?

The point I am trying desperately to make in the face of much anticipated skepticism is that we can be just as proud of 'giving birth' to a new silicon-based form of life as we can be giving birth to baby humans. There is simply no logical reason to want to give birth to a biological baby in preference to a *better* silicon-brained baby, unless we deem the replicating patterns of genes to be somehow better than the replicating patterns of memes. They are both just patterns with the ability to direct energy toward their own replication. The genes are no more 'our' replicators than are such memes as, say, those of 'our' native language. It takes both genes and memes to construct the kinds of humans we strive to be.

We humans are defined as much or more by memes as we are by genes. And, our noble educational endeavors are dedicated to acquiring an amount of memes that far exceeds the number of our inherited genes. The memes account for nearly all of the difference between our closest animal cousins and ourselves. So, why should we root for genes over memes? Both are absolutely vital to us, and we should want to propagate our memes to our children just as much as we should want to propagate our genes. We should be just as proud of giving birth to memetic inventions as to genetic babies. In both cases – genes and memes – we are talking about information, not matter. It is the information in the patterns of our genes and the blueprints of our memes that replicates, evolves and fulfills our definition for successful life.

I don't see anything wrong with giving birth to a new form of silicon based life through the development of computers and robotics, while at the same time continuing our biological tradition of having babies. We should be proud to have the opportunity and the capability of so doing. Our goal should be to create intelligence, better than ours, that can combine memes in new ways, leading to new and important discoveries. Such is the clear path to societal survival. Of course, I am playing right into the hands of the memes when I suggest this, but that's okay, because I see valid memes and the understanding of them as being as close to God as one can get. In fact, whatever concept anyone ever holds as their highest moral authority, that concept, if it can be remembered and communicated to others, must be a meme. Therefore, memes will always define our morality. The complete understanding of all valid memes must be our higher calling.

The trend toward ever greater numbers of memes over genes may be an inevitable consequence of evolution's arrow. After all, there is no doubt that we humans are relying more and more on the memetic designs of computers to make our lives more convenient. Already, there are dozens of computer chips in a single car to make it safer, more comfortable, and more efficient. And, businesses use computers and robotics to make production more accurate, more reliable and more efficient. Already, the pentagon uses computers for surveillance, for battlefield simulation and to aid in quick, accurate decision making.

Now, let's extrapolate this trend to a world in which nearly all tasks are handled by computers, and the role of humans is merely to keep the computers running. When we reach the point at which the operation of computers becomes more important than human life, won't we then have become the slaves, and the computers our masters? When our role is merely to keep the computers running, then I'm afraid we will serve their interests even though we will be handsomely rewarded with safe and convenient lives. Could that time be much more than fifty years away? At some point it will become more important to keep the computers running than it will be to preserve a human life. I'm sure our military already holds that priority. In fact, we may already have become slaves to our memetic technology.

If a transition from biological life to machine life is inevitable, we would do better to embrace it. Remember that, over the long term, we can only push the evolution of life or technology in the direction it is statistically destined to go – along the arrow of evolution.

In the opening pages of this book, I cited Ray Kurzweil's summarization of technology's acceleration. He convincingly shows that computers today are approaching a level of computational capability comparable to the brain of a mouse. In 2030, trends put them at the capability of a human brain, and in 2060, a single computer can be expected to have a level of computing power exceeding the combination of all brains of all humans.

A skeptic might gleefully point out that my projection is based on the assumption that recent trends will continue. That same skeptic might further assert that, in fact, we are approaching some fundamental limits on the advancement of computing technology. Such a skeptic would be absolutely right; however, the same argument has been made time and time again, just before many of the breakthroughs that have allowed technology to overcome previous limitations. For example, the power of calculating technology has progressed smoothly through five different substrates over the past hundred years: mechanical devices, relays, vacuum tubes, discrete transistors, and finally, integrated circuits. It almost seems as though nature was designed to enable these smooth transitions.

Am I allowed to make the argument that computational limitations tend to succumb to new technologies, and that new technologies will surely allow us to break through the hard limits we see today? Am I allowed to make the argument on the basis of the mere hundred years worth of evidence we have, regarding computing technology? It would be nice to have a little bit more data. Well, we do have quite a bit more data. When we broaden our view of computational technology to include even biological technology, we see a much bigger trend of acceleration that has been in place for billions of years. To see this, we must take a step back and recognize that technology is not a human accomplishment, it is an evolutionary accomplishment – a statistically inevitable product of deterministic processes. When we take such a generalized memetic perspective, it becomes apparent that significant technological *events* have been happening ever since life began to direct energy toward its own replication.

In Kurzweil's book *The Singularity Is Near* (2005), he plots a graph of the significant events of innovation, not just by humans, but also by evolution in general. He starts with the initial event of the creation of life, followed by the emergence of multicellular organisms, then the Cambrian explosion, then reptiles, then mammals, and so on, all the way through to the technological events created by humans, ending with: electrical devices, computers and, finally, personal computers. His graph plots the time between events as a function of when they occurred. A plot in logarithmic time shows a very clear straight line, which indicates acceleration in normal time. This is indeed significant evidence toward the idea that there is an inevitability to the progress of natural events. They have been coming ever faster at a constant rate of acceleration for billions of years. We would be pretty arrogant to think we could somehow alter nature's inevitable progress. We would be silly to assume it will end soon.

Kurzweil's use of the term *singularity* in the title of his book refers to a coming moment in time. He establishes from his plot of historical data that the moment will take place around the year 2045. Just as the mathematical function $1/(2045-t)$ gets bigger and bigger at an ever faster rate, until, at midnight on new year's eve of the year 2044, it explodes to infinity, so will there be a period of time, just before the singularity – just before the big crystal ball touches the ground in Times Square – when new events come so fast that there will be almost zero time between them. In the time it takes for the big ball to drop, an extremely intelligent computer, that was itself designed by a slightly less intelligent computer, will find a cure for cancer, discover a way to postpone old-age, figure out how to defy gravity, calculate how to travel faster than light, and, well ... who knows what else. Allow me to analogize this strange concept.

Physicists speak of a singularity at the center of a 'black hole' – a collapsed star whose gravitational field, produced during the process of its collapsing, is so enormous as to cause the collapse to continue forever. Theoretically, it gets smaller and smaller forever, and, therefore, exists only as a single point in space. The singularity in technological time that we are fast approaching is like the singularity at the center of a black hole. As we get closer to it, we go ever faster. But instead of gravity doing the sucking, pulling ever harder as we get closer, it is technology that pushes us ever harder at an increasingly faster rate.

Black holes are funny things. They bend and stretch time and space. If we were to carry a yardstick while falling toward a black hole, it's length would change as we got closer. And time would slow as we got faster. But, these strange effects would only be apparent to a stationary observer. The meaning of time and space is relative to an observer's frame of reference (that's why Einstein's theories are called 'theories of *relativity*'). This is amazing! It means that my universe is slightly different than your universe, although they are definitely connected.

In a loosely analogous manner, our technological universes vary from person to person, from culture to culture. You know things that I don't know, and I know things that you don't know. It means that one of us is closer to the coming technological singularity than the other. This effect will certainly become more pronounced as we approach the singularity, just as the warping of time and space become more pronounced with the approach toward a black hole. Yet, the technological differences among us may not make any more of a difference to anyone than the tiny relativistic differences between our perspectives on space and time. On the other hand, as we approach the technological singularity, our slight differences may determine who lives and who dies. The people who control the amazingly powerful computer that discovers all the as-yet-undiscovered knowledge will hold a huge advantage over everyone else.

Kurzweil sees life on the other side of the singularity as being omniscient – no more disease, no more suffering – just perfect knowledge of medicine, agriculture, energy production, deep-space travel, and everything else. Although, he also admits to the possibility, vehemently espoused by Sun

Microsystems' former chief scientist Bill Joy, of self-destruction due to such sophisticated knowledge of weaponry and nanotechnology.

Bill Joy warns of several possible catastrophes, among them, a 'gray-goo' scenario that goes something like this: Suppose nanotechnology allows us to make molecular-sized robots capable of self-replication, and also capable of such miraculous feats as, say, metabolizing the production of hydrogen from sunlight and water. Our energy problems would be over, but other, worse, consequent situations might be lurking right around the corner. The problem comes from the possibility of random mutations to the self-replicating 'nanobots' yielding the potential for creating little critters that, instead of metabolizing just sunlight and water, can also metabolize all biological matter. They would spread quickly across the surface of the world, leaving nothing but the metabolic by-product of *gray goo* in their wake.

Like the futility of looking backward to what might have existed before the 'big bang' – another singularity in time – we are having trouble seeing forward, to a time after the technological singularity. Here is a relevant quote copied from Kurzweil's book (2005, p. 22):

> We humans have the ability to internalize the world and conduct "what if's" in our heads; we can solve many problems thousand of times faster than natural selection. Now, by creating the means to execute those simulations at much higher speeds, we are entering a regime as radically different from our human past as we humans are from the lower animals. From the human point of view, this change will be a throwing away of all the previous rules, perhaps in the blink of an eye, an exponential runaway beyond any hope of control.
>
> **– Vernor Vinge**, *The Technological Singularity* (1993)

The trend of technology clearly points to an amazing transformation, coming soon. Yet, I can certainly sympathize with a skeptical attitude. From our daily perspectives, there is a stability to our experiences that lead us to believe that tomorrow will be just like yesterday. But, on the other hand, it seems inevitable that self-replicating memetic patterns of technology – intelligent robots giving 'birth' to more intelligent robots – will emerge well before we reach the technological singularity. And, indeed, such an emergence can be expected to facilitate the singularity. It will likely happen within this century, possibly within the first half of this century. This is a pretty weird line of thought. I find it inevitable, yet at the same time, very non-intuitive. Let us put it aside and step back onto firmer ground.

The Intelligence of Gaia

Imagine taking a giant step backward and viewing earthly life from a much more distant perspective. From such a vantage point we begin to see an intelligence greater than human intelligence. Soon, we'll take an even more distant perspective, from a cosmological viewpoint, and we'll see the possibility of yet an even higher level of intelligence operating on the very creation of universes. For now, we'll limit our scope to planetary intelligence.

When life on a planet is viewed from a distant perspective, the salient feature is the degree to which replicating patterns of life cooperate with each other in the general process of directing energy toward their mutual replication. Such is the essence of a phenomenon James Lovelock recognized while working at NASA on projects concerned with developing the means for detecting life on distant planets. Lovelock's ideas are expressed in his book *Gaia* (1979).

Lovelock realized that, in a planetary environment, the fluid components (oceans and atmosphere) will largely be responsible for managing the metabolic ingredients and the output products of whatever life exists there. This is quite evident in the homeostatic nature of our earthly environment. The levels of gaseous products of life, such as methane, oxygen and carbon dioxide, seek a point of equilibrium that balances the effects of life against nature's tendency toward chemical equilibrium. Surprisingly, the levels seem to intelligently adjust toward maximizing the amount of life that can possibly exist in the given environment. We should expect the same to be true for any instance of planetary life.

Consider the changing environment that any planet is likely to witness during the ongoing development of life. Gaseous products of life will tend to collect in the atmosphere until reaching levels harmful to the aggregate life that produces them. But, as the environment changes from the products of life, new forms of life are made possible. And, if any of those new forms of life happen to be fully complementary to the earlier forms of life, they will enable both new and old forms of life to flourish. They then cooperate toward their mutual perpetuation. I'll restate these principles in a real-world example.

The earthly presence of oxygen originated as a toxic waste product of cyanobacteria. As the level of oxygen grew in the atmosphere, it drove many then existing species to extinction. But, since then, new forms of life have emerged that thrive on oxygen. Indeed, it now enables the metabolism of humans and many other animals. And, of course, our human acts of consuming oxygen and producing carbon dioxide cooperate nicely with the complementary acts, performed by plants, of consuming carbon dioxide and producing oxygen. The current arrangement between plants and many animals, including humans, is nicely complementary and can be considered as cooperative. But it is important to realize that the cooperative arrangement occurred as a result of an earlier act that caused a regression in the well-being of aggregate life.

The homeostasis of life in the aggregate makes the Earth appear as something of a single living organism that itself probes various cooperative states and thereby evolves so as to maximize its own aggregate well-being, just as all living species do. After all, it seems silly to say that life adapted to Earth's environment, when in fact life largely produced the environment. The proper way to look at it is to say that life and the environment co-evolved. And, amazingly, that co-evolution itself appears to be an intelligent process tending toward maximizing life's potential, in the aggregate.

Lynn Margulis makes a compelling case, along similar lines, that we humans live on a *Symbiotic Planet* (1998). She is a bold thinking person who has finally received widespread credit for having proposed the now-commonly-accepted idea that the mitochondria found in modern cells came about long ago through the symbiotic merging of bacteria. She further suggests that many evolutionary advances can be explained as a merging of symbiotic DNA. Margulis makes a good case that all the scientific evidence of life supports Lovelock's theory of Gaia – the idea that all earthly life, taken in aggregate, tends to change its own environment in a way that is ultimately conducive to the furthering of life. This is another example of nature's "incessant compulsion for self-organization."

There are indeed reciprocal and circular dependencies among many species of earthly replicators that can best be described as cooperative arrangements. Just as an economy composed of many cooperatively interacting companies appears as a single, huge, conglomerate company at a higher level, so does the principle of cooperation among many species of planetary life yield a single aggregate instance of life at a higher level. The binding agent is always cooperation. Indeed, nature gives selection preference to things that cooperate, simply because cooperation yields synergy. In addition to the natural selection of individuals that cooperate with each other, we now find evidence that nature even prefers species that cooperate. That preference is revealed through the principle of Gaia, and is completely consistent with the notion of cooperationism (previously discussed in chapters two and four).

Indeed, our human genes cooperate beautifully with the genes of many other species of life. If the degree of success for a species is measured by the existing number of active copies of related genes – and I can't think of a better metric – then our human genes cooperate nicely with, say, the genes of pigs. As Robert Wright points out, even though we raise pigs for slaughter, we cause their numbers to thrive in the process. We effectively guarantee the perpetuation of the pig species. From the perspective of an individual pig, its relationship to humans cannot be characterized as cooperative, but from the perspective of its perpetually replicating genes, the view is quite different – very beneficial.

We may attribute the tremendous success of human genes to their highly cooperative strategies of 'partnering' with the genes from many other species of life. Human genes 'use' the phenotypic products of many species, and in turn, foster the perpetuation of those species. Stated more succinctly, our genes have, for a long time, 'fit' well with the environment. Sadly, however, we humans are now actively destroying millions of species of organism that we may later find to be critical to the delicate balance of earthly life. We'll return to this concern in a moment.

The principle of Gaia implies a sort of intelligence for maintaining the environment at a point that is conducive to life in the aggregate. Such a belief is quite sensible from my viewpoint. But, Lovelock and Margulis have been ridiculed for their theories of Gaia by many, including the likes of Richard Dawkins. He says the following:

> Lovelock rightly regards homeostatic self-regulation as one of the characteristic activities of living organisms, and this leads him to the daring hypothesis that the whole Earth is equivalent to a single living organism. Whereas Thomas's (1974) likening of the world to a living cell can be accepted as a throwaway poetic line, Lovelock clearly takes his Earth/organism comparison seriously enough to devote a whole book to it. He really means it.

– **Richard Dawkins** (1982) *The Extended Phenotype*, p.235

Dawkins, being a strong proponent of biological evolution, apparently expects an organism of Gaia to evolve in a manner that is strictly analogous to the style of evolution with which he is most familiar. He seems to expect all evolutionary processes to evolve in a manner identical to the manner in which biological life evolves. So, he improperly characterizes Lovelock's Gaia hypothesis as requiring "a set of rival Gaias, presumably on different planets," most of which would have to be "dead planets whose homeostatic regulation systems had failed," yet among them would be "a handful of successful, well-regulated planets of which Earth is one" (*The Extended Phenotype*, p.236).

I believe Dawkins is making a mistake here. It is particularly important that we recognize this mistake, because many cognitive scientists make a very similar mistake when they are confronted with the Darwinian view of cognition, which is, after all, the central theme of this book. Since they can't find strict analogies to biological Darwinism, they dismiss it as ridiculous. But, it turns out that there are many variations on the theme of evolution. A true believer in biological evolution should not find it hard to accept that indeed the process of evolution itself can evolve.

The mistake made by Dawkins and others results from their failure to recognize a serial, as opposed to parallel, style of evolution. Allow me to present a view of Gaia, conveniently dismissed by Dawkins as unworkable, in which the evolution of homeostatic states can occur on a single planet. Here's how:

Earthly life clearly changes with each generation of its inhabitants. Considered in aggregate, life on Earth constantly mutates into something new. Mutated patterns combine in various ways to explore a space of homeostatic states ranging from simple, independent patterns to more complex, cooperating patterns having greater efficiency of directing energy toward ever faster exploration of various states. The result is a unified earthly organism that mutates ever faster, as time goes by.

Now, think back to a previous analogy I made with regard to how patterns of neural firings evolve in a serial fashion. If we use a complex maze to analogize the various paths evolution can possibly take, then, to loosely analogize the way biological evolution explores the maze is to suppose we put many mice in all at once. Mice that take different paths represent different species of life. Dawkins seems to be stuck on the idea that evolution needs many parallel variants, like many different mice, all exploring the maze simultaneously. But, realize that a single mouse can still intelligently explore the entire maze, and will eventually find the cheese, by backtracking whenever it reaches a dead-end. Thus, the single mouse can analogize a serial process of evolution. This is how we can view the evolution of Gaia. Aggregate life goes down a particular path guided by the parallel evolution of each species, but, occasionally, it may have to back up and try a different path, corresponding to a different combination of species.

One might infer that the serial style of exploration by Gaian evolution seems to preclude the elements of competition and selection. Yet, I suggest there *is* an element of selection occurring in a different manner than what we consider to be typical. Gaian evolution searches for a system of *stability*. When it finds one, it tends to persist. Stable systems are thereby *selected* over unstable systems. Because of the temporal persistence of stable states, we may say that nature prefers them over unstable states. Indeed, if we ever find life on a distant planet, it will be more likely to be in a relatively stable state than in an extremely unstable, dynamic state. We can expect it to have previously gone through many unstable dynamic states, but dynamic systems always tend toward stability, and once they get to a stable state, they tend to stay there simply by virtue of the stability. The more stable a system is, the longer it will likely stay that way, guaranteed simply by definition of the word 'stable'.

Recall my analogy for simulated annealing: Put a 'Mexican jumping bean' on a hilly landscape, and watch as it hops around, sampling various coordinates, sliding down slopes into various local minima. If the bean looses energy at a slow enough rate, it will be guaranteed to find, and stay in, the global minimum (corresponding to the optimum point) of the landscape. This too is an intelligent process that operates in a serial manner. Parallel exploration by multiple jumping beans would discover better states faster, but a single bean can get the job done.

We may mathematically define the landscape such that its global minimum corresponds to the global maximum of stability in some system. So, when the bean finally settles into the global minimum, it will have found the most stable state. As the bean hops around, there is a corresponding competition between states that are represented by local minima. For example, let us imagine the bean hopping between two valleys separated by a hump. The lower of the two valleys, corresponding to a higher stability, will always win the competition, if indeed the energy of the bean reduces gradually enough. So, a serially mutating system *can* exhibit competition and selection between various mutations on the basis of stability. Now, let us try to interpret how it might be happening in the evolution of Earth's environment.

Plants rely on animals for carbon dioxide and animals rely on plants for oxygen. The mutually satisfying relationship results in a relatively stable level of both oxygen and carbon dioxide in the atmosphere. And because of that homeostasis, life continues to thrive and evolve. If homeostasis were to be violated, say, by some mutant plant that randomly discovers a way to save itself the costs of manufacturing oxygen, then (in Dawkins' words) "it would outreproduce its more public-spirited colleagues, and genes for public-spiritedness would soon disappear." No doubt, a strong violation of public-spiritedness by a rampantly reproducing species could upset the delicate balance. But, in a mutually beneficial relationship, if one side doesn't hold up its end of the bargain, then both sides will ultimately suffer. In Dawkins' example, a selfish plant that thrives by saving itself the costs of manufacturing oxygen might eventually eliminate all oxygen and thereby destroy the animal kingdom, but in so doing, would destroy its own source of carbon dioxide.

Indeed, a violator of Gaia might cause the whole system to suffer a severe setback. The whole system must then go back to an earlier state and try again, like a single mouse retracing its steps after reaching a dead-end.

Consider the emergence of a devious defector in a group of cooperators. Acts of defection cause the group as a whole to suffer. Likewise, an entire Gaian organism can suffer from a temporarily successful defecting renegade replicating element. But such a setback will only open up the opportunity for the emergence of other sorts of cooperative elements that may be less susceptible to the emergence of such renegade defection. And, in fact, sometimes the effect of defection can be locally contained. A form of life that strongly violates Gaia can be like a disease that kills its victims too quickly, effectively destroying its own means for propagation. It can snuff itself out in its own local environment.

The evolution of life in the aggregate will occasionally fall back to an earlier state and then try again to find a new and different path through homeostatic state space. But, while an instance of planetary life might occasionally regress to some earlier state, it need not die to evolve. When aggregate life finds a point of stability, which will likely involve many diverse elements cooperating through various loops of interaction, it will tend to remain there. And, just as distributed computing tends to be more stable than centralized computing, so do diversely cooperating elements provide a measure of stability by diversifying risk.

Perhaps, in fact, Earth did witness many such attempts at the violation of Gaia. In some cases, perhaps, the entire system regressed. But, whatever happened, a cooperative system appears to have ultimately emerged, having great stability in its methods of cooperation. Just as a single mouse will eventually find the cheese in a maze, so can a single planetary system, through fits and starts, eventually happen upon a stable system of diversely distributed types of replicating elements all cooperating together toward their mutual perpetuation. Nature apparently seeks an impervious system of stability, which is most likely to be composed of many diverse and cooperating components.

Perhaps the Gaian situation in which we find our planet is the result of earthly life having previously bailed out of many situations where Gaia was easy to violate. Perhaps we don't see plants violating Gaia by saving themselves the costs of producing oxygen because earthly life has evolved into a homeostatic state such that the laws of nature prevent such a plant from possibly existing. Given that the stability of various cooperative systems is completely determined by the laws of nature, perhaps earthly life has discovered the optimally stable arrangement, defined by natural laws. Or, perhaps, Earth will yet suffer more major setbacks.

While the Gaian network of Earth is absolutely built of self-interested components, it is not hard to see that self-interest for any replicating pattern always involves an interest in a stable and conducive environment. Such an environment will necessarily include other types of stable patterns with which a given self-interested pattern can cooperate. So, one of the most selfish things a replicator can do is to cooperate with many other replicators.

To see more clearly how self-interest and cooperation are compatible and encouraged by nature, let us consider an interesting thought experiment, called *Daisyworld*, proposed by Lovelock. The experiment demonstrates exactly how a system built of self-interested components can achieve a regulated environment that is good for every individual component.

Imagine a world covered with white and black daisies, and further imagine that the daisies thrive within a certain temperature range, but die otherwise. Black daisies absorb sunlight and therefore do better at the colder end of the range. White daisies reflect sunlight and therefore do better at the warmer end of the range. Black daisies, in the process of absorbing sunlight, tend to warm the environment. White daisies, in the process of reflecting sunlight, tend to cool the environment. Now, with this common-sense background knowledge, let us start the experiment by supposing a cold planet covered by a preponderance of black daisies.

The black daisies absorb the light, convert it to heat, and heat up the environment. They will thrive until the environment gets near the hot end of the comfortable temperature range. At that point, white daises will do better than black daisies. As white daisies flourish, they will reflect sunlight and thereby cool the environment. An equilibrium temperature will eventually maintain the relative numbers of black and white daisies at some stable value. Nature seeks stability.

Now, if the sun were to suddenly increase its luminosity, for whatever reason, the black daisies will convert the increase in sunlight to an increase in heat, and the environmental temperature will rise. But, recall that white daisies thrive better than black daisies in warmer temperatures, and so, the white daisies will flourish, thereby increasing their proportion of the total daisy population. The increase in white daisies will lower the environmental temperature toward the optimal value. Thus, the environment is regulated by self-interested components. They just happen to be configured in a way that naturally allows them to cooperate together toward their mutual benefit. And such mutual benefit rewards them all with the enhanced likelihood of perpetuating their defining replicators.

> ... Gaian patterns appear to be planned but occur in the absence of any central "head" or "brain." ... Without any extraneous assumptions, without sex or evolution, without mystical presuppositions of planetary consciousness, the daisies of Daisyworld cool their world despite the warming sun.
>
> – **Lynn Margulis** (1998) *Symbiotic Planet*, p.126

The appearance of planning, to which Margulis refers, gives the further appearance of intent, purpose and intelligence. Such is the nature of even human intelligence. While human intelligence has the appearance of purpose, such purpose is always to perpetuate the replicators that define it. And, while human intelligence has the appearance of creative intent, such creativity and proactive planning happens merely through a fortuitous organization of neurons that enables an evolution of firing patterns operating within a simulated environment.

The intelligent characteristics of Daisyworld simply result from the fortuitous existence of automatically replicating devices that are able to cooperate toward altering the environment in a way that is conducive to their mutual perpetuation. And if such a fortuitous situation did not exist, then the aggregate life would be more likely to fail, regress, and have to try again. It is the nature of aggregate life that it automatically tries again and again until a fortuitous situation of stability is discovered. Thus, fortuitous situations become inevitable.

I see nothing wrong with considering earthly life in the context of Gaia. The best environment for humans simply must be one that is also good for all the various forms of life with which we humans cooperate, and on whom we humans depend. We may indeed consider ourselves to be part of a larger life. I don't see how Dawkins can so strongly object to the characterization of life on Earth as a single sort of organism until he clearly defines exactly what life is.

I find that Dawkins' writing always assumes a freedom of human will that can only be explained through a sort of supernatural essence, even though he strongly rejects such a soul. For instance, when he says that "We, alone on earth, can rebel against the tyranny of the selfish replicators" (*The Selfish Gene*, p.201), he is implying a freedom of will that acts independently of the selfish replicators – the genes and memes. But, from where would such a will emanate? When scientists and philosophers finally come to accept that all minds are deterministic, then the only definition for life that makes any sense at all is one in which interdependent and cooperating patterns are to be considered as parts of the same instance of life, no matter their physical proximity.

Now, here is an interesting twist that I believe shows exactly why we must think of ourselves as cooperative parts of a larger life. Our brains are like computers that automatically do what they are programmed to do until an external event causes them to shift to a different programmed activity. You may cause such an event that is external to my programming (by sending me email, for example), thereby shifting my thoughts from their otherwise determined path, and I may cause such an event that is external to your programming (by replying), thereby shifting your thoughts from their otherwise determined path. But, without any external influences, your thoughts and my thoughts will proceed down paths that are determined by physical laws operating on the neurons of our respective brains.

So, one of the stranger aspects of our having deterministic minds is that I am able to affect your thoughts and you are able to affect my thoughts, but I am not able to affect my own thoughts and neither are you able to affect your own thoughts. How then can we possibly view interacting life to be anything other than an interconnected web? Let us face up to it – humans deterministically pursue goals that have been genetically programmed into them, by way of plans that have been memetically programmed into them. And, it is the laws of nature that shape those evolving genes and memes so as to encourage them toward the creation of loving, caring, sharing, compassionate and cooperative beings.

I find it odd that Dawkins and many of his disciples adamantly reject any sort of volitional free will at the level of godly creativity (biological evolution), yet insist on it at the level of human creativity (thinking). If the ability to transcend deterministic causation through free will and conscious creativity exists at all, shouldn't we prefer to imagine such an ability existing at the godly level before the human level? I believe the only scientific position that makes any sense is one that assumes there is no such thing as volitional conscious creativity, at any level. There is only the process of evolution operating at all levels, with the accompanying *belief* in conscious feelings of free will at the level of humans.

Whereas the concept of Gaia was initially met with a great deal of skepticism and criticism by many, it has more recently been gaining acceptance under a new name – Earth System Science. So, perhaps it was the name itself that garnered contempt. Indeed, the word 'Gaia' is Greek for 'Earth Goddess' or 'Mother Earth'. And, as such, the name implies an essence of agency, as if the aggregate organism of Earth has a sort of consciousness and a corresponding volitional intent to keep itself balanced. But, Gaia involves no more volitional agency than does the mammalian ability to automatically maintain a stable body temperature, or even the human ability to automatically find likely paths to future happiness. Gaia simply occurs when many species settle into an acceptable and stable environmental arrangement.

The homeostasis of Gaia is a good thing for all cooperating patterns of life, and it is far more than just poetry to describe the world, in the sort of pop ecology terms that Dawkins apparently deplores, as "a fine-meshed network of interrelationships, a web of communications which it has taken thousands of years to build up, and woe betide mankind if we tear it down" (*The Extended Phenotype*, p.237). Earth's environment is very stable due to the extremely wide diversity of species, but humans now have the technological means for upsetting the stability by destroying huge numbers of species. Once the stability is gone, it could take eons to re-stabilize.

Gaia is a natural consequence of nature's preference for life that cooperates. If I were to sum up nature's morality in one line it would have to be: Either find a way to cooperate with the successful elements of Gaia, or get out of the way.

Intelligent Design vs. Evolution

There has been a recent resurgence of debate between evolutionists and creationists, with the battle now taking place over the theory of *Intelligent Design*. Creationists have recently adapted their stance. Recognizing that all scientific evidence is in strong support of evolution, some creationists have chosen to reconcile their views with that evidence. So, they have moved the locus of their divine creation from the unlikely Garden of Eden to the gaps in evolutionary evidence. They hold the unsupported belief that God accounts for the discrete jumps in fossil evidence. Such a belief allows them to continue proclaiming that only an intelligent God could have designed the intricate complexity and wonderful morality of human life. This claiming of new high-ground by creationists has inspired some evolutionists, including Richard Dawkins, to spend a great deal of time and energy arguing that biological design by nature is nothing more than an illusion of intelligence. But then, according to the theory of mind presented here, so must intelligent design by humans be just an illusion.

Frankly, I have to laugh at both sides in the heated debate. Both sides are blissfully unaware of the fundamental truth that renders the argument absolutely meaningless. A mind is creative by way of the very same sort of process that biological evolution is creative. There is little or no distinction to be made between the two processes. The concept of a Darwinian style of cognition, as developed here, makes perfectly clear that intelligent design and evolution are, in their essence, the very same thing.

Of course, if I am forced to take sides, my beliefs are much more in line with most evolutionists. Indeed, even after the overly-complex theory of Intelligent Design has been adjusted to be in alignment with scientific evidence, it still carries a lot of extra fat eligible to be cut away by Ockham's razor. Evolution is perfectly capable of having bridged the gaps in fossil evidence, and hence, is a much simpler theory than one that includes a meddling sort of god.

Pure and universal evolution is clearly the simplest and best theory, but many evolutionists tend to ignore its obvious implications. For instance, many of them wrongly believe that there is some sort of essence in human intelligence that can't be captured by a computational process. Indeed, many of them wrongly suppose a freedom of will in human choice. And, many of them completely overlook the majestic complexity made possible by the perfect balancing of universal constants. Such a precise tuning implies a contemplative tuner.

Also wrong are those evolutionists, epitomized by Stephen J. Gould, who have suggested that the variability associated with evolution is so random that, if the last three billion years could be replayed, the resulting life would be far different from what we find on Earth today. My earlier discussion of evolution's arrow makes clear that there is a statistical destiny for all life. Such a universal destiny even includes moral behavior.

Perhaps the proper view of life is one that embraces both the mysticism of religion *and* the logic of evolution. So, allow me to move the locus of divine creation to the design of evolution's magnificent arrow. If indeed there is an arrow of evolutionary development along which all life will necessarily progress, it is then logical to ask: Who or what designed the magnificent arrow pointing the way toward the eventual evolutionary emergence of intelligent and moral life? The only answer we have is: God. But such a style of god is unlikely to be an anthropomorphic style who hears and answers our prayers. It is more of a mathematical formula and associated 'boundary condition' responsible for the origin of everything. But, whatever you call it, God's creative thought process – biological evolution – seems to be largely indistinguishable from man's creative thought process – Darwinian cognition. There is now some common ground on which religionists and evolutionists may stand, and hold each others' hands.

English philosopher William Paley once suggested that a watch found in the wilderness would quickly be identified as a product of intelligence, not of nature. He means to imply that all life, like the watch in the wilderness, clearly shows similar indications of intelligent design. But such a stance seems to presuppose a creative process that transcends natural capabilities – a spiritual sort. When one realizes that the human mind operates by deterministically manipulating patterns, the distinction between creative thought and biological evolution largely disappears.

Just as we quickly recognize William Paley's watch, laying on a mossy forest floor, as a product of intelligent design, so must we also recognize all life as a product of intelligent design. But the design process is merely an iterated form of evolution in either case. Whereas Paley's watch was designed by a Lamarckian style of evolution, biological life, so far, has evolved by a non-Lamarckian process. But, that will soon change with the advent of genetic engineering. We will eventually be able to reverse-engineer a desirable trait back to its responsible genes, and then incorporate those genes in future humans. Genetic evolution will then also be Lamarckian.

We can better appreciate the similarity between human intelligence and the intelligence of evolution by considering how life will eventually evolve in the future. As genetic engineering becomes commonplace, we will inevitably be faced with 'designing' future generations of humanity. The evolution of life will then occur by intelligent genetic modification. But, how will we do it? What characteristics would we as a society consider to be good? If we were to legislate for or against certain genetic characteristics, what would they be?

Should we prefer, for instance, for our children to be taller or shorter than we are? To answer that question, we might imagine a world full of people who are eight feet tall, or a world full of people who are four feet tall. Well, it would take more food to sustain a population of tall people than to sustain a population of short people, but food is plentiful here in the U. S., so perhaps that isn't a matter of concern. But think about the average height of doorways and ceilings in important buildings. Are we prepared to tear down and rebuild our existing structures to accommodate our tall descendants?

Notice that the intelligent process is one of trial and evaluation whether it is performed in the intelligent human mind or by evolution. We imagine a world full of tall people and evaluate how well they might fare. We imagine them eating more food and stooping as they walk through doorways. If our imaginations allow us to conceive of our descendants as prospering from being taller, then we might feel justified in engineering people to be taller. Indeed, evolution has been 'trying out' tall people in the real environment for eons. If they do better, then genetics evolve so as to make people taller.

From biology to technology, there seem to be a great many intelligent processes at work in the world, but at their roots, for each and every one of them, the intelligence amounts to nothing more than automatic trial, evaluation and selection. Even intelligent computer software automatically achieves its goals by evaluating many different trials within a computationally simulated environment.

The lowest levels of evolution tend to generate trial runs via random mutations. As higher levels of intelligence emerge from the designs produced at lower levels, they tend to generate their trial runs in a more organized fashion. Random mutations yield to recombinant modules. Non-Lamarckian systems eventually discover ways to reverse-engineer things so as to produce more efficient (more intelligent) Lamarckian processes. And serial processes of evolution yield to competitive parallel processes, which are simply able to explore more trial runs in less time. It appears that the Darwinian process of trial, variation, and selection by evaluation is responsible for all intelligence. And it also appears that the process of evolution is itself destined to evolve in predetermined ways, becoming ever faster and more efficient.

Allow me to finish out the book by taking a short diversion into some wildly speculative concepts. My hope is that the curious mind will find this diversion to be somewhat entertaining and mildly interesting.

Beyond Universal Darwinism

Suppose that some sort of evolutionary process is at the root of human intelligence, and that even our human wills are guided by an automatic Darwinian process operating toward the satisfaction of genetically defined goals. Suppose, as I have suggested several times, there is no source of creativity anywhere in the universe other than the process of evolution. Such a universal Darwinian perspective has allowed us to understand the true nature of the human characteristics we hold so dear – morality and intelligence. So, it is quite appropriate that we now look to the god-like Darwinian process as a source for the sort of creativity that might have been responsible for intelligently designing the entire universe, along with all the associated laws of nature that establish and determine the magnificent arrow of evolution.

Let us extend Darwinism beyond being universal, to include an evolutionary process operating even above the level of the universe. Reconsider our brief discussion of the anthropic principle (in chapter four), in which it was suggested that the natural constants of our universe appear to be finely tuned for enabling human-like life. Even the slightest deviation of some universal constants, from what they happen to be, would have rendered the emergence of intelligent life impossible, or so the speculation goes.

Now, recall (from chapter one) that data only becomes *information* when it provides value to an instance of life. And also recall that the Copenhagen interpretation of quantum dynamics, in order to force the collapse of possible quantum states down to a single state, requires that sentient life participate by interaction or mere observation. It seems that the universe is intimately concerned with life, and nothing else. Here is how Kurzweil puts it: "The rules of our universe and the balance of the physical constants that govern the interaction of basic forces are so exquisitely, delicately, and exactly appropriate for the codification and evolution of information (resulting in increased complexity) that one wonders how such an extraordinarily unlikely situation came about." (Kurzweil, 2005, p.15).

It seems that the constants of nature were designed to be what they are. But, designed by whom or by what? This is indeed welcome support to religionists who wish to claim that the universe and all its laws are the products of Intelligent Design by some sort of benevolent god. But, it equally supports the position of universal Darwinists who speculate that universal constants may have evolved in much the same way that life has evolved. From my persepctive, the creative mind of God and the Darwinian process of intelligent design are beginning to merge into a single sort of phenomenon.

Imagine that our universe is simply one of many, and that every universe can somehow spawn baby universes having slightly different physical laws and constants. Then, perhaps our universe is the descendent product of a string of ancestral universes, each of whose constants were slightly modified from the generation before. Lee Smolin, in his book *The Life of the Cosmos* (1997), proposes just such an evolution of cosmological constants by supposing that 'baby universes' are the natural products of black holes. If it is indeed the case that black holes create baby universes with constants that are able to vary slightly from those of the universe in which they are spawned, then universes with constants that encourage black hole production will give forth more progeny and thereby become more plentiful over the generations. And, the resulting variations among the many possible universes may allow some of them to be life-friendly.

In a book titled *Biocosm* (2003), James Gardner takes this idea one step further by including intelligent life into the evolutionary loop. He suggests that universes having constants conducive to the emergence of intelligent life will become more plentiful if life is able to become intelligent enough to engineer the greater production of black holes, and hence, even more baby universes. If it is possible for highly intelligent life to design the constants of baby universes, through the engineering of black holes, such designs could, with each generation, produce cosmological constants that are ever more precisely tuned to support intelligent life.

Unfortunately, this theory fails to explain how intelligent life came about in the first place. The very first universe – the mother of all other universes – must have already been life-friendly enough to generate highly intelligent life. It must have already been at the place where the evolution of universes promises to take us. Gardner's theory essentially requires that we start with the sort of intelligent life that we are relying on cosmic evolution to explain. Perhaps we can assume that Smolin's simpler mechanism of evolving universes 'bootstrapped' the creation of intelligent life. Yet, the cosmological constants don't seem to have evolved toward the greater production of black holes. Or, perhaps, as Gardner supposes, the curvature of time allows life in the future of our universe to design our universe of the past. This explanation falls well short of being scientifically satisfying, in my mind. It still asks us to suppose the 'pre-existence' of what we are trying to explain, even though its existence somehow springs from the future.

Another problem with Gardner's theory has to do with motive. Why should instances of intelligent life want to create baby universes without the possibility of benefiting from them? Further, as a problem for both Smolin's and Gardner's theories, we don't know with any confidence that black holes actually do produce baby universes. And even if they do, we don't know of any way that cosmological constants in those baby universes might be specified by the intelligent cosmic engineers. While Gardner's view is pleasant to think about, especially for speculating that life has a noble and selfless purpose, the theory has a lot of explaining yet to do.

It is indeed *emotionally* satisfying to believe that we humans will eventually discover how to create baby universes having laws that we choose, and that our morality is evolving toward a selfless sort of benevolence that would cause us to want to create beings in our own image, with no hope of ever benefiting from their existence. That would clearly put us in a godly role. But perhaps there is a more scientifically satisfying explanation for the creation of baby universes – one that does not require cosmic engineering performed by highly intelligent life. Indeed, I'll propose a very simple and plausible explanation showing quite clearly that billions of baby universes are birthing and dying as you read this.

To understand this bold concept we need to free our minds from the ruts in which our thoughts tend to travel. We must go back to the fundamental things we know. A universe, so far as we can tell, is just a place where bits of matter and energy 'exist' and obey certain mathematical rules that define how they interact and move around in space. The current arrangement of all atoms in our universe results from the strict mathematical rules of physics having operated on their mathematical locations and velocities since the beginning of time.

Any quantum randomness that may or may not exist in our universe seems pretty imperceptible at levels we can measure. So, rifle bullets never make U-turns in mid air; people who jump off of burning buildings never fall upward; and, rarely can we notice perceptible differences between maternal twins, even though they result from completely independent and extremely complicated processes of gene expression, operating over many years on identical genetic instructions. Nature's extremely high reliability, at the level of scale that we typically experience, allows us to think of nature's rules as being purely mathematical formulas, with perhaps a tiny bit of randomness thrown in – a very tiny bit. Now, let us contemplate the birth of just such a mathematically determined universe.

I remember seeing as a kid some fun little toy things – small gray rocks. Just add a drop of water to one and it quickly grows in size by more than an order of magnitude, mushrooming in all directions. Allow me to use this simple mental image as a visual metaphor describing the auspicious start of a universe. To build yourself a universe, first, get a singularity and then add a 'drop' of mathematical laws. Sit back and watch as it grows by many orders of magnitude every second, strictly according to those universal laws.

Perhaps you won't even need a singularity; perhaps the mathematical laws themselves *predict* the singularity at time t=0. In that case, all you need to do is 'endorse' the set of mathematical laws, as is done inside a computer when it calculates the beautiful pattern of a fractal image. The mere act of endorsing, or enabling, a proper set of cosmological laws may be enough to cause virtual matter and energy to spring into mathematical 'existence'. Matter that is described mathematically must be virtually 'real' to any other mathematical entities with which it computationally interacts. In fact, any forms of life that might eventually emerge in a mathematically described universe would have no way of detecting whether the matter and energy they sense are actually real or simply computationally simulated.

We may even think of our own reality as something of a huge parallel computer, in which each point of space continuously calculates its next state on the basis of its own current state and the states of its neighboring points in space. This characteristic of our universe allows us to represent it computationally. Our brains apparently do just that when they simulate our environments. And we are programming our computers to do so with even far greater precision.

To better understand this concept of mathematically defined reality, think of a computer that simulates the weather conditions of our environment. It does so by mathematically modeling the physical interactions between big blocks of the atmosphere in order to predict how those blocks will change their temperature, pressure, humidity, and air flow, over time. For example, we know that if the air pressure in one block of air is higher than in a neighboring block, there will be a force of wind moving air along the pressure gradient. Heat similarly flows along temperature gradients. All we need do is represent various blocks of air with numbers representing pressure, temperature, humidity and wind velocity. We already know how to mathematically describe the forces of meteorology and their effects, at least, approximately. So, as we mathematically apply those forces, over and over again, the numbers change – the system comes alive.

Such a 3-dimensional weather simulator is crudely approximated for us, visually, when we watch the weather report. Looking at the weather map is like looking at a 2-dimensional version of the full 3-dimensional simulator. If we were to lay a grid over the weather map, we could quickly learn to model how cells of weather affect each other. We would soon discover that the prevailing winds of the 'jet stream' usually cause weather patterns to drift eastward (in the northern hemisphere). We would also find that localized winds usually tend to blow from regions of high pressure toward regions of low pressure. Perhaps we could find a set of mathematical formulas that, when put into each and every cell, would accurately predict the various attributes of temperature, pressure, humidity, and so on. That is exactly what meteorologists try to do.

There has been a lot of research devoted to computational simulation and prediction, for the general purpose of predicting the behaviors of many sorts of natural systems. In one such field of study, known as *cellular automata*, a simple formula is used to compute the behavior of every cell in a grid. For every moment of time, the formula, when applied at any given cell, simply takes as input the previous moment's outputs in the same cell and surrounding cells. For certain governing formulas, the grid can produce amazingly complex and dynamic patterns when seeded and repetitively calculated, again and again, over every cell.

If it turns out to be the case that we can accurately model all of nature by some sort of cellular automaton, then we might wonder if perhaps nature actually *is* a cellular automaton. Some brilliant minds have speculated as much. For example, the renowned physicist Richard Feynman wondered about modeling nature as a cellular automaton after his careful consideration of the relationship between information, matter and energy. And the brilliant mathematician Norbert Weiner, in his book *Cybernetics* (1948), suggested that the transformation of information, not energy, represents the fundamental building block of the universe.

Stephen Wolfram, in his book titled *A New Kind of Science* (2002), explores the amazing complexity of cellular automata resulting from the recursive iteration of very simple formulas, according to very simple rules. He suggests that a greater understanding of these simple replicating systems is essential to the advancement of science in many of the areas that have proven to be intractable by our traditional tools of mathematics and logic. One need only look at the beautifully organized complexity of a fractal image to appreciate that the appearance of magnificent design can naturally arise from simple mathematical formulas repetitively iterated over billions of cycles.

Wolfram apparently got the idea for using cellular automata to model nature from Edward Fredkin, a brilliant inventor who served on the faculty of MIT even though he never graduated college. Fredkin's most notable contribution to science is his theory of *digital physics*, which states that the universe is not continuous, but instead is composed of discrete elements. He suggests that time and space are broken into little cells that act as if their behaviors are calculated by a machine. Along with Konrad Zuse (who originally proposed a rough version of the theory), Fredkin has been searching for the rules by which a very finely meshed 3-dimensional cellular automaton could accurately model the universe. "You see, I don't believe there are objects like electrons and photons, and things which are themselves and nothing else ... What I believe is that there's an information process, and the bits, when they're in certain configurations, behave like the thing we call the electron, or the hydrogen atom, or whatever." (Fredkin, quoted by Robert Wright, 1988)

Fredkin's theory is related to the famous Church-Turing thesis (by Alonzo Church and Alan Turing), stating that all mathematical functions conceivable by the human mind can also be calculated on a very simple computational device. Alan Turing (1936) proposed such a computational device, known as the *Universal Turing Machine*, which laid the foundation for John von Neumann's conceptualization of the modern computer (von Neumann also invented the study of cellular automata). Given that a modern computer has been proven to be capable of calculating absolutely any conceivable mathematical function, it logically follows that such a computer would be able to calculate all the laws of nature, whatever they may be. Indeed, Fredkin firmly believes: "what cannot be programmed, cannot be physics."

So, now, imagine a very powerful computer that simulates physical interactions down to the atomic level. Suppose we could simulate the natural laws of motion, momentum, and effects of forces on atomic particles. Suppose we could simulate atomic binding, molecular folding, and even quantum effects, all within a simulated environment that includes the simulated existence of necessary ingredients for cellular processes. With such a theoretical capability, a computer would be able to simulate the interactions of DNA molecules and all the molecules that allow DNA to replicate. Let's put a copy of the human genome into the simulator and see what happens. Watch, as proteins are expressed by the mathematical rules of physical laws; watch as the process of mitosis causes cells to divide in the development of an embryo; and, after some virtual time, watch the functioning of the completed body and even the brain.

Let us continue to extend the power of our simulator by including in the simulation all the physical properties of an earth-like environment. Suppose we watch as the simulator carries out all the physical interactions of matter and energy enabling the functions of human life, including human thought. Would the simulated human have *feelings* of consciousness as we real humans have? As I have previously stated, very plainly: If it doesn't have *feelings* of consciousness, then the simulator is simply not accurate. If the simulation of all the physical elements in the virtual human's brain is completely accurate, then the simulated life absolutely must experience consciousness, exactly as we humans do.

It seems there can be no discernible difference between what we perceive as our 'real' universe and the perception by virtual life of its own virtual universe, which is actually simulated through continuous mathematical iteration within some sort of a computational device. We are thereby forced to entertain the idea that our universe may, in actuality, be simulated by some sort of hellishly powerful computer. I am not so interested in claiming that our universe is simulated as I am interested in claiming that mathematically simulated universes can seem just as 'real' to life inside them as our universe seems 'real' to us. But perhaps the best way to make my case is to suppose that our universe *is* simulated. So, let us imagine an enormously powerful computer able to mathematically model the virtual state of an entire universe such as the one in which we exist.

Imagine that every particle of matter and every photon of energy in our universe is merely represented by a set of numbers in some computer that iterates time by continuously calculating the next incremental cosmological state – say, a femtosecond later – from the current state. In such a simulation, the position, the momentum, the spin and all other physical characteristics of every subatomic particle constituting a physical body amount to nothing more than a set of numbers undergoing transmogrification by strict mathematical rules.

Matter need not exist as hard, touchable stuff, except in the perceptions of simulated minds. Instead, particles of matter can exist merely as numbers, and the same is true for bits of energy. But virtual matter does become hard, touchable stuff by way of a computational rule enforcing a law that prevents two bits of matter from occupying the same space at the same time. The law is mathematically expressed by a repulsive force that grows stronger as two bits of matter are brought closer together. The computational enforcement of that law causes virtual matter to become virtually hard and touchable, as the act of bringing one's virtual fingers in close proximity with some virtual object results in the virtual sensation of the object pushing back through the repulsive force.

Is it ridiculous to consider that we humans might be existing inside the simulated world of some very powerful mind or computer, similar to the way humans are depicted in the popular movie *The Matrix*? That movie portrays humans as existing inside individual life-support pods, with their brains connected to a central computer that coordinates all their thoughts and perceptions in some virtual world that it creates. But we are taking the idea one step further by eliminating the physical bodies altogether, leaving only a simulated world filled with simulated people.

The idea of our real world being merely simulated seems rather far-fetched, don't you think? It would take a pretty big and powerful computer to simulate the kind of detailed universe we seem to experience. And we haven't yet come close to building a computer capable of such computational power, not even within a few orders of magnitude. But, computers have only existed for about fifty years, during which they have increased in power by well over a full order of magnitude every decade. The next fifty years promise even more incredible advances than we've seen so far. So, let us merely hold out the very remote possibility that it might someday be feasible to simulate an entire universe, but much smaller than our own. Now, if we accept this as a possibility, then some interesting philosophical consequences begin to emerge.

The most profound consequence goes to James Gardner's effort to put intelligent life inside the iterated loop by which universes might evolve. It turns out that intelligent life need not *intend* to spawn baby universes through the cosmic engineering of black holes, as Gardner suggests. Indeed, the spawning of baby universes need not involve black holes at all. Instead, the very process of intelligent thought already spawns baby universes through mental simulation, automatically and inadvertently.

Realize that, as intelligence naturally evolves, it does so through more accurate and detailed simulation. One species is more intelligent than another if its mental simulation of reality is more accurate than the other's. And, one individual is more intelligent than another if it has learned more about how the world works than the other has learned, thereby allowing it to better simulate the world. Scientific knowledge is the acquired ability to accurately simulate the physical world, mentally. It is what allows us to conceive of new inventions during the process of mental simulation.

The arrow of evolution seems to guarantee that minds, whether real or artificial, will evolve toward ever greater powers of simulation. Throughout this book, I have argued that intelligence requires simulation of the environment. It should now be obvious that the fittest forms of life – the most intelligent – are those that can best simulate the real world in their imaginations for the purpose of prediction. Indeed, we humans routinely assess our various opportunities for investment and cooperation by predicting the various outcomes through mental simulation.

The incredible implication of all this is that our human thoughts actually create baby universes. Those universes aren't precisely the same as the one in which we 'exist', but they are simulations of small portions of it, nonetheless. This idea takes my evolving thoughts in all kinds of weird directions. For example, when I imagine a frog sitting in my hand, perhaps that frog is 'alive' in some limited sense and in a fuzzy sort of virtual world of my thoughts, for that brief moment. As I continue to think about that frog, it seems to hop from my hand of its own volition, sometimes landing on its back, sometimes on its face, unlike a cat that I might imagine holding in my arms, which always lands on its feet when it jumps.

My imagination actually seems to hold an all-possible-worlds set of scenarios, in which the frog simultaneously jumps *and* sits in my hand at the same time. This is reminiscent of how the Copenhagen interpretation of quantum dynamics demands that all quantum possibilities actually occur simultaneously, until an event of observer participancy forces them to collapse down to a single state. Only when I think about the frog interacting with someone or something does my mind have to choose between the many possible things the frog might do.

Simulated life must often engage in recursive simulation, in much the same manner that I mentally simulate the thinking of my wife when I buy her a present – it is the only way I can predict whether or not she'll like it. Through my occasional mental simulation of her, I can even imagine my wife simulating me in her thoughts as she decides what to get me for a present; it is the only way I can drop subtle hints as to what I want, without her realizing it.

As another example, I might contemplate what my wife's reaction will be if I tell her of some important fact about which I know. I may need to consider her reaction before I decide to tell her. But then, my simulation of her must simulate her thoughts, including all the simulations that her brain might perform. Embedded simulations create universes within universes, none of them existing within the same space-time because all versions of space-time are merely mathematically simulated.

While my wife might be simulated in my brain, both my wife and I might be simulated in the brain of a higher form of life, and *that* instance of life might be simulated in a still higher form of life, and so on. Additionally, various simulated universes can be connected by something like wormholes through which information may flow. For example, my wife's thoughts do certainly share much information with my thoughts. Information flows between our mentally simulated universes whenever we speak to each other.

At each level of simulation, a new universe is spawned, having characteristics similar to its parent universe. This is just the sort of recursive mechanism that could allow cosmological constants to evolve through generations of descendent universes. And, we must always keep in mind that an extrapolation of the current course of evolution, from what we see here on Earth, almost guarantees that embedded and recursive simulations will take place in extremely powerful computers sometime in the near future. Indeed, our universe seems to be intent on developing intelligent life that is able to simulate the world in its fullest detail. This is in complete accordance with the idea that our universe is a product of an evolutionary process of the sort James Gardner sought, which would require intelligent life to be inside the evolutionary loop in order to facilitate it.

It seems that simulated universes have been temporarily birthing and dying in the minds of individuals for as long as there has been intelligent life. It is the very nature of all intelligence to create computational descriptions of virtual realities in which to assess various possible plans of action. And now, like never before, computers are often creating temporary universes through simulations of mathematical worlds. There is no question that the future of accurate simulation belongs to computers. We will increasingly rely on ever more sophisticated computer models to simulate the processes of life. As we get ever more accurate in the simulation of human life, then that simulated life must eventually become self aware. If it doesn't, then the simulation is simply not yet accurate. But realize that any sort of awareness can only apply within the context of its simulated environment. It cannot know it is simulated unless we program the simulator to tell it, although it very well might discover on its own enough science to suspect it is simulated.

By projecting the evolution of intelligent life on Earth far into the future, it becomes inevitable that simulators will gain significant power over time. Simulation provides the ability to predict the future for various scenarios, and the ability to predict the future is a tremendous survival advantage. Life will inevitably evolve to the point where it develops computational minds that are nearly perfect simulators. At that point, it absolutely must be the case that living entities, similar to humans, will virtually exist inside computationally simulated universes.

So, just how plausible is the idea that universes evolve by spawning new universes through mental simulation? Well, my father thinks this idea is just plain wacky. And, recognizing that he is a very intelligent man, I have to admit he is absolutely right. It is wacky. But, the fact that we and our incredible universe exist at all is simply wacky to begin with. And so, any explanation that we come up with, for how and why we humans came to exist, is likely to be wackier still. I feel vindicated by the fact that, like most critics of the theory, my father is unable to propose a less wacky theory. Indeed, there is no other simple explanation, outside of those theories that have been thoroughly soaked in the spiritual magic of religion.

If we use Ockham's razor to assess the wackiness of various explanations regarding the origin of our universe, we find that the *simulated universes* explanation is not wacky at all. It is scientifically parsimonious for the following reasons, all of which we can be quite certain: (1) Intelligence enhances fitness. (2) Intelligence requires mental simulation. (3) Life always evolves toward ever greater intelligence, which requires ever more accurate and powerful mental simulation. (4) Life existing inside a simulator can't know whether or not it is simulated. (5) Very sophisticated forms of life will eventually be simulated by very powerful computers on our planet in the not too distant future. (6) If we ever discover cosmological constants that breed life more quickly than the constants we find in our own universe, we will likely use those life-friendlier constants in some of our simulations.

In fact, we will probably, at some point in time, run many simulations of the beginnings of our universe using many slightly different constants, just to see if there are constants that are more life-friendly. I consider it to be nearly *inevitable*, then, that we will replicate our universe many times, through simulation, with occasional mutations.

Whether or not we humans are merely simulated, there will almost certainly be simulated humans existing in virtual universes sometime in the future. And I see no reason why those mathematical humans can't eventually be every bit as conscious and 'self aware', within the context of their own computed environments, as we are. All we need to do is accurately simulate the workings of our brains.

Whether or not such powerful simulation actually *has* happened (in the creation of our universe), we can be quite certain it eventually *will* happen, which makes it quite plausible, in the absence of anything better, as an explanation for what *has* happened.

Earlier, I criticized Gardner's model of evolving universes as suffering from a problem of infinite regress. Even though we can easily imagine an ancestral universe to our own that might have been slightly less life-friendly, we can't extend that line of reasoning all the way back to some original universe having no life at all. Gardner's model, involving life's engineering of black holes, relies on the pre-existence of intelligent life in order to facilitate the evolution of universal constants necessary for *creating* intelligent life. Does my model suffer from the same problem of infinite regress?

The critical question is whether a simulated universe of complexity N can create an embedded universe of complexity greater than N, for all values of N. If so, then we have sufficient means for cosmological constants to evolve all the way from a very bland original universe to the complex style of life-friendly universe in which we live. The original universe – the mother of all universes – only needed to develop enough intelligence to crudely simulate its crude environment. And, one of the many crude simulations running in that original universe could have produced a new environment slightly more similar to the one we enjoy.

Perhaps the mother of all universes was just sophisticated enough to allow the emergence of replicating patterns having the ability to intelligently model the environment in a very rudimentary way. For instance, a single bit in a pattern can model the presence or absence of light in its environment. And such a rudimentary model might indeed cause the respective pattern to better survive and replicate contingent upon the knowledge represented in that bit. Several bits can model shades of light. Hundreds of bits can begin to model the dimensions of space. Each model represents an embedded, self-contained, albeit extremely rudimentary, universe. But we need to conceive of the means for generating a simulated universe of higher complexity, which requires huge numbers of bits all cooperating toward the process of modeling. Is this possible?

Indeed, it just may be possible for a very primitive mental simulation to produce a more life-friendly universe than the one in which it exists. Just as our human intelligence now seems capable of producing a style of computer intelligence that will someday exceed our own, so perhaps can a primitive universe spawn a simulated baby universe having cosmological constants that are slightly more life-friendly than its own.

Consider an analogy. A computer having 1 Megabyte of fast non-persistent memory (RAM) and 100 Megabytes of slow persistent memory (disk) can use the inherent locality of reference in typical software to appear as though it has 100 Megabytes of virtual memory that is both fast *and* persistent. It simply uses the 1 Megabyte of fast memory as a cache for the much larger slow memory. In a similar manner, perhaps a universe can make use of some natural shortcuts to simulate embedded universes having greater virtual complexity than its own.

Or, perhaps a huge but cosmologically simple universe can trade some of its space or time or matter or energy in exchange for added complexity of cosmological constants in the construction of simulated baby universes. In fact, we know such trade-offs are possible in our universe. We routinely program our computers so that they apply very complex formulas to small amounts of data. Thus, we seem to be able to dedicate lots of matter and energy in our universe toward the simulation of a more complex, but much smaller, simulated universe. Realize also that virtual time depends on speed of computation, which can be increased through parallelism. So, it may be possible to simulate the entire life of a smaller universe in a fraction of the life of its larger parent universe. This points to a fundamental trade-off allowing the exchange of simple redundant matter for time and complexity in the process of simulation.

Perhaps, in addition to the trade-off between matter and energy (captured by Einstein's formula $E=mc^2$), there is a more fundamental relationship between matter and space-time, when traded from one universe to a simulated child universe. We need only imagine that our ancestral universes are bigger in time and space, or have more stuff in them, even though they may be less complex than our own. When we extrapolate all the way back to the original mother universe, it need only be extremely large (as if ours isn't large enough) with just enough complexity to: (1) allow patterns to automatically and differentially replicate themselves, and (2) allow portions of those patterns to reflect (simulate) some characteristics of their respective environments.

So, what does this all mean? Well, in reference to the anthropic principle, Freeman Dyson once wrote, "The more I examine the universe and study the details of its architecture, the more evidence I find that the universe must have known that we were coming." That comment now makes some plausible sense under the theory that our universe is the result of automatic and deterministic simulation by similar life in some other universe, and that we experimental protégés may exist solely for the purpose of exploring various possibilities toward enhancing that other life's ability to perpetuate its own defining replicating patterns. But how can we ever know for sure?

We have to wonder whether there might be some clues that would allow virtual life to realize it is simulated within some sort of a computing device. For example, perhaps 'garbage collection' routines (memory reclamation and allocation processes) inside a simulating computer would appear as black holes to simulated life within that computer – returning unused resources to the available pool of free, allocable resources. Or, in another line of reasoning, perhaps the inability for a computer to precisely know its own complete internal state would manifest itself as something like Heisenberg's uncertainty principle to life that is simulated within that computer. Or, perhaps uncertainty relates to the finite precision, and consequent round-off error, of floating-point numbers typically used in computer simulations.

Perhaps the gradual reduction of energy in the optimizing algorithm of *simulated annealing* would manifest itself as something like steadily rising entropy to life that is simulated within a computer. In fact, the amount of energy in our universe that is usable for thermodynamic work is indeed inversely related to entropy. So as entropy steadily and inevitably rises, the usable energy in our universe gradually declines. This is consistent with the idea that our universe was created as a simulated environment for the purpose of optimizing some parameter through a process like annealing.

There is still more circumstantial evidence from quantum physics that is consistent with our universe being simulated. It is best described by a popular thought experiment known as *Schrödinger's Cat*. In a box is placed a cat and an apparatus for killing the cat on the basis of some random event possibly occurring within the next hour. The apparatus might consist of, say, a piece of radioactive material having a 50% chance of emitting a product of decay in the next hour, and a radioactive detector linked to a canister of poison gas. Upon detection of decay, the gas is released, which kills the cat.

According to the Copenhagen interpretation of quantum mechanics, we must assume that the cat, having a 50% chance of being killed during the hour, actually exists in both states, dead *and* alive, just before the end of the hour. Only when the box is opened and examined by some observer does nature make up her mind as to the state in which the cat actually exists. This odd characteristic of nature is the only explanation of physics that satisfies all the experimental results of quantum-level experimentation. It caused Einstein to defiantly exclaim: "Do you really think the moon isn't there if you aren't looking at it?" Niels Bohr replied: "Einstein, don't tell God what to do."

Does this strange phenomenon give us any clues as to whether our universe is a product of computational simulation? I believe it does represent interesting circumstantial evidence: Perhaps the parsimonious use of compute cycles by an efficient programmer causes anomalies at the quantum level of a virtual universe to appear as though virtual nature doesn't actually decide between possible outcomes of an event until it absolutely has to, as a consequence of virtual observer participancy.

As a programmer, myself, I know that I have taken similar shortcuts in the past. I often structure my programs to perform calculations only when necessary. After all, if the purpose of a simulated universe is to assess the behavior of life, then why should it bother to compute the reflection of light from the moon if no one is looking at it. Why should it bother to calculate the status of Schrödinger's cat while the cat cannot affect the state of anything else in the universe?

Why indeed? As Stephen Hawking once asked (1988, p.174): "Why does the universe go to all the bother of existing?" The answer to that question can only be attributable to what we typically refer to as God. In the simulated-worlds scenario, our information will always be restricted to our own universe, unless our 'hacker' creators decide to reveal the truth to us by way of, say, programming the simulator to write flaming letters across the sky.

Are there any circumstances that would compel them to do such a thing? The answer is clear when we consider what we humans would do with life that we might eventually create inside a computer simulator. If we, for example, try to predict what life will be like on the other side of the technological singularity, by simulating it, why would we ever feel a need to let such a simulated life know its purpose? Such a revelation would ruin the experiment. And besides, that simulated life would be nothing more than just bits of information, destined to dissipate into nothingness as soon as we pull the plug and go home for the evening. Indeed, so might we humans be nothing more than just bits of information, all the while, believing we are something much more substantial and even spiritual.

Regarding the revelation of ultimate purpose, it seems likely that circumstantial evidence is all we can ever hope for. But fortunately, all the circumstantial evidence of science points to a clear and beneficent natural purpose for life: to advance intelligence and to promote cooperative morality among all replicating entities in whatever ways possible. That is what makes life better in absolutely any environment.

Bibliography

Aunger, Robert (2002) *The Electric Meme*. New York: The Free Press.

Axelrod, Robert (1984) *The Evolution of Cooperation*. New York: Basic Books.

Axelrod, Robert and Hamilton, William (1981) "The Evolution of Cooperation." *Science*, vol. 211, pp. 1390-1396.

Bateson, P. P. G. (1978) "Book review of *The Selfish Gene*." *Animal Behavior*, Vol. 26.

Blackmore, Susan (1999) *The Meme Machine*. Oxford: Oxford University Press.

Blackmore, Susan (2002) "Why Does Consciousness Only Seem to Exist When You Look For It?" *New Scientist*, June 22.

Blackmore, Susan (2004) *Consciousness: An Introduction*. New York: Oxford University Press.

Cairns-Smith, A. G. (1985) *Seven Clues to the Origin of Life*. Cambridge: Cambridge University Press.

Chaisson, Eric (2001) *Cosmic Evolution*. Cambridge: Harvard University Press.

Chomsky, Noam (1975) *Reflections on Language*. New York: Pantheon.

Darwin, Charles (1859) *The Origin of Species*. London: John Murray.

Dawkins, Richard (1976a, 1989) *The Selfish Gene*. Oxford: Oxford University Press.

Dawkins, Richard (1976b) Hierarchical Organisation : A Candidate Principle for Ethology. In *Growing Points in Ethology* (eds P. P. G. Bateson & R. A. Hinde), pp. 7-54. Cambridge: Cambridge University Press.

Dawkins, Richard (1982) *The Extended Phenotype*. Oxford: Oxford University Press.

Dawkins, Richard (1987, 1996) *The Blind Watchmaker*. New York: Norton.

Dawkins, Richard (1995) *River Out of Eden*. New York: Basic Books.

Dawkins, Richard (1998) *Unweaving the Rainbow*. New York: Mariner.

Deacon, Terrence W. (1997) *The Symbolic Species*. New York: Norton.

Dennett, Daniel C. (1984) *Elbow Room: The Varieties of Free Will Worth Wanting*. Cambridge: MIT Press.

Dennett, Daniel C. (1991) *Consciousness Explained*. Boston: Little, Brown.

Dennett, Daniel C. (1995) *Darwin's Dangerous Idea: Evolution and the Meanings of Life*. New York: Touchstone.

Dennett, Daniel C. (2003) *Freedom Evolves*. New York: Penguin.

Descartes, Rene (1641) *Meditations on First Philosophy*. Paris, trans. 1901 by John Veitch.

Distin, Kate (2005) *The Selfish Meme*. Cambridge: Cambridge University Press.

Edelman, Gerald M. (1987) *Neural Darwinism: The Theory of Neuronal Group Selection*. New York: Basic Books.

Edelman, Gerald M. and Tononi, Giulio (2000) *A Universe of Consciousness: How Matter Becomes Imagination*. New York: Basic Books.

Ferris, Timothy (1997) *The Whole Shebang*. New York: Touchstone.

Gärdenfors, Peter (2000) *Conceptual Spaces: The Geometry of Thought*. Cambridge: MIT Press.

Gardner, James N. (2003) *Biocosm*. Hawaii: Inner Ocean Publishing.

Gleick, James (1987) *Chaos: Making a New Science*. New York: Penguin.

Grey Walter, W. (1963) Presentation to the Osler Society, Oxford University.

Haeckel, Ernst (1899) *Riddle of the Universe at the Close of the Nineteenth Century*. New York: Harper and Brothers, trans. 1900 by Joseph McCabe.

Hawking, Stephen W. (1988) *A Brief History of Time*. Bantam Books.

Hawkins, Jeff and Blakeslee, Sandra (2004) *On Intelligence*. New York: Times Books.

Hofstadter, Douglas (1999) *Godel, Escher and Bach: An Eternal*

Golden Braid. New York: Basic Books.

Holland, John (1995) *Hidden Order: How Adaptation Builds Complexity*. New York: Perseus Books.

Holland, John H.; Holyoak, Keith J.; Nisbett, Richard E.; Thagard, Paul R. (1986) *Induction: Processe of Inference, Learning and Discovery*. MIT Press.Hawkins

Kolers, P. A. and von Grünau, M. (1976) "Shape and Color in Apparent Motion," *Vision Research*, vol. 16, pp. 329-335.

Kurzweil, Ray (1999) *The Age of Spiritual Machines: When Computers Exceed Human Intelligence*. New York: Viking.

Kurzweil, Ray (2005) *The Singularity Is Near: When Humans Transcend Biology*. New York: Penguin.

Libet, B. (1981) "The Experimental Evidence for Subjective Referral of a Sensory Experience Backwards in Time: Reply to P. S. Churchland," *Philosophy of Science*, vol. 48, pp. 182-197.

Libet, B. (1985) "Unconscious Cerebral Initiative and the Role of Conscious Will in Voluntary Action," *Behavioral and Brain Sciences*, vol. 8, pp. 529-566.

Lorenz, Edward (1963) "Deterministic Nonperiodic Flow," Journal of the Atmospheric Sciences.

Lovelock, James (1979) *Gaia*. Oxford: Oxford University Press.

Margulis, Lynn (1998) *Symbiotic Planet*. New York: Basic Books.

Maynard Smith, J. (1982) *Evolution and the Theory of Games*. Cambridge: Cambridge University Press.

McCune, W. (1997) "Solution of the Robbins Problem," *Journal of Automated Reasoning*, 19(3), pp. 263-276.

Minsky, Marvin (1988) *Society of Mind*. New York: Simon & Schuster.

Moore, G. E. (1903) *Principia Ethica*. Cambridge: Cambridge University Press.

Moore, G. E. (1942) "Reply to My Critics" *The Philosophy of G. E. Moore*. (Edited by P. A. Schilpp) Evanston Illinois: Northwestern University Press. p.582.

Mountcastle, Vernon (1978) "An Organizing Principle for Cerebral Function: The Unit Model and the Distributed System" *The Mindful*

Brain. (Gerald M. Edelman and Vernon B. Mountcastle, eds.) Cambridge: MIT Press.

O'Reilly, Randall C. and Munakata, Yuko (2000) *Computational Explorations in Cognitive Neuroscience.* MIT Press.

Penrose, Roger (1989) *The Emperor's New Mind: Concerning Computers, Minds, and the Laws of Physics.* Oxford: Oxford University Press.

Pinker, S. (1997) *How the Mind Works.* New York: W. W. Norton & Company.

Pinker, S. (2002) *The Blank Slate: The Modern Denial of Human Nature.* New York: Penguin.

Plotkin, Henry C. (1993) *Darwin Machines and the Nature of Knowledge.* London: Penguin.

Rand, Ayn (1957) *Atlas Shrugged.* New York: Penguin.

Rawls, John (1971) *A Theory of Justice.* Cambridge: Harvard University Press.

Rees, Martin (2000) *Just Six Numbers: The Deep Forces That Shape the Universe.* New York: Basic Books.

Ridley, Matt (2003) *Nature via Nurture: Genes, Experience, and What Makes Us Human.* New York: Harper Collins.

Rosenthal, D. (1986) "Two Concepts of Consciousness," *Philosophical Studies*, 49, pp. 329-359.

Rosenthal, D. (1989) "Thinking That One Thinks," ZIF Report No. 11, Research Group on Mind and Brain, Perspectives in Theoretical Psychology and the Philosophy of Mind, Zentrum für Interdizsiplinäre Forschung, Bielefeld, Germany.

Rosenthal, D. (1990a) "Why Are Verbally Expressed Thoughts Conscious?" ZIF Report No. 32, Zentrum für Interdizsiplinäre Forschung, Bielefeld, Germany.

Rosenthal, D. (1990b) "A Theory of Consciousness," ZIF Report No. 40, Zentrum für Interdizsiplinäre Forschung, Bielefeld, Germany.

Russel, Stuart and Norvig, Peter (2003) *Artificial Intelligence: A Modern Approach.* New Jersey: Pearson Education.

Sagan, Carl (1977) *The Dragons of Eden: Speculations on the Evolution of Human Intelligence*. New York: Random House.

Searle, John (1980) "Minds, Brains and Programs," *Behavioral and Brain Sciences*, vol. 3, pp. 63-108.

Sheehan, Evan Louis (2005) *The Laughing Genes: A Scientific Perspective on Ethics and Morality*. Bloomington, Indiana: AuthorHouse.

Shermer, Michael (2004) *The Science of Good and Evil*. New York: Times Books.

Sigmund, Fehr and Nowak (2002) "The Economics of Fair Play," *Scientific American*, January, vol. 286, pp. 82-87.

Simon, Herbert (1962) "The Architecture of Complexity," *Proceedings of the American Philosophical Society*, December 106:6, pp. 467-82.

Skinner, B. F. (1971) *Beyond Freedom and Dignity*. New York: Knopf.

Smolin, Lee (1997) *The Life of the Cosmos*. Oxford: Oxford University Press.

Spencer, Herbert (1879) *The Data of Ethics*. Edinburgh: Williams and Norgate.

Strogatz, Steven (2003) *Sync: The Emerging Science of Spontaneous Order*. New York: Hyperion Books.

Surowiecki, James (2004) *The Wisdom of Crowds*. New York: Doubleday.

Trivers, Robert (1985) *Social Evolution*. Menlo Park, Calif.:Benjamin/ Cummings.

Turing, Alan (1936) "On Computable Numbers, With an Application to the Entscheidungsproblem," *Proceedings of the London Mathematical Society*, Series 2 Volume 42.

Turing, Alan (1950) "Computing Machinery and Intelligence," *Mind*, vol. 59, pp. 433-460.

Waldrop, M. Mitchell (1992) *Complexity*. New York: Simon and Schuster.

Wegner, Daniel M. (2002) *The Illusion of Conscious Will*. Cambridge: MIT Press.

Wiener, Norbert (1948, 1965) *Cybernetics: or Control and Communication in the Animal and the Machine*. Cambridge: MIT

Press.

Wilson, Edward O. (1975) *Sociobiology: The New Synthesis.* Cambridge: Harvard University Press.

Wilson, Edward O. (1978) *On Human Nature.* Cambridge: Harvard University Press.

Wilson, Edward O. (1998) *Consilience: The Unity of Knowledge.* New York: Vintage Books.

Wolfram, Stephen (2002) *A New Kind of Science.* Wolfram Media, Inc.

Wright, Robert (1988) "Did the Universe Just Happen?" *Atlantic Monthly* (April, pp.29-44) http://digitalphysics.org/Publications/Wri88a/html.

Wright, Robert (1994) *The Moral Animal: The New Science of Evolutionary Psychology.* New York: Pantheon Books.

Wright, Robert (2000) *Nonzero: The Logic of Human Destiny.* New York: Vintage Books.

Index

—

Endnotes

[1] A Necker Cube is a simple stick figure of a cube. It can easily be drawn by first drawing two overlapping and parallel squares, square A and square B, and then drawing lines connecting the respective corners from square A to square B. Depending on your perspective – that is, depending on which square, A or B, you believe to be in front – the cube takes on different orientations.

[2] Stenograph machines have been used for many decades by court reporters to record spoken testimony. By expressing words phonetically, each of the 24 keys expresses more information than any key of a typewriter, and allows transcription at speeds up to 225 words per minute.

[3] Prisoner's Dilemma is a game that reveals the benefits of cooperation, even in a completely computational environment. For more details, see *The Laughing Genes*.

[4] Julia Child – a well-known chef who specialized in French cooking and appeared on television quite often.

[5] I won't ever try to hide the fact that I am just a man with biological urges. In fact, I make no attempt at writing with gender neutrality because I believe it is important for the reader to be constantly reminded that this is written from a male perspective.

[6] The *naturalistic fallacy*, defined by G. E. Moore in his book Principia Ethica (1903), declares that moral values cannot be derived from anything natural. The idea is similar to philosopher David Hume's suggestion that *is* does not imply *ought*.

[7] Behaviors are *inclusively adaptive* when they aid in the propagation of one's own gene patterns, including the identical copies of those gene patterns that are likely to exist in siblings or offspring.

[8] A grilled cheese sandwich once sold for $28,000 on eBay simply because it appeared to have an image of the Virgin Mary grilled onto it. And, I once saw a news report stating that thousands of people had journeyed to see an apparent image of the Virgin Mary in the reflection of sunlight from a particular dirty plate-glass window.

[9] Physicists have proposed a concept known as 'string theory' suggesting that reality is actually composed of at least ten dimensions, most of which can only be detected mathematically. Is there any connection between the many dimensions of reality and of conceptual space? Maybe. I'll later present some compelling evidence that the characteristics of the universe should be similar to the simulation characteristics of highly advanced mental processes.

About the Author

With credentials that include a doctorate in electrical engineering, a masters degree in business administration, work experience at IBM's prestigious T. J. Watson research headquarters, and the successful entrepreneurial development of an Internet gaming platform, the author, who writes under the pen name Evan Louis Sheehan, is currently working at Inteplex, Inc. on the design of a computer architecture intended to explore various computational aspects of human-like thinking. He lives in Delaware with his wife and two children, and also serves on the board of directors for a company that develops genetic sequencing techniques for microbe identification.

Visit: TheMockingMemes.com for more information.